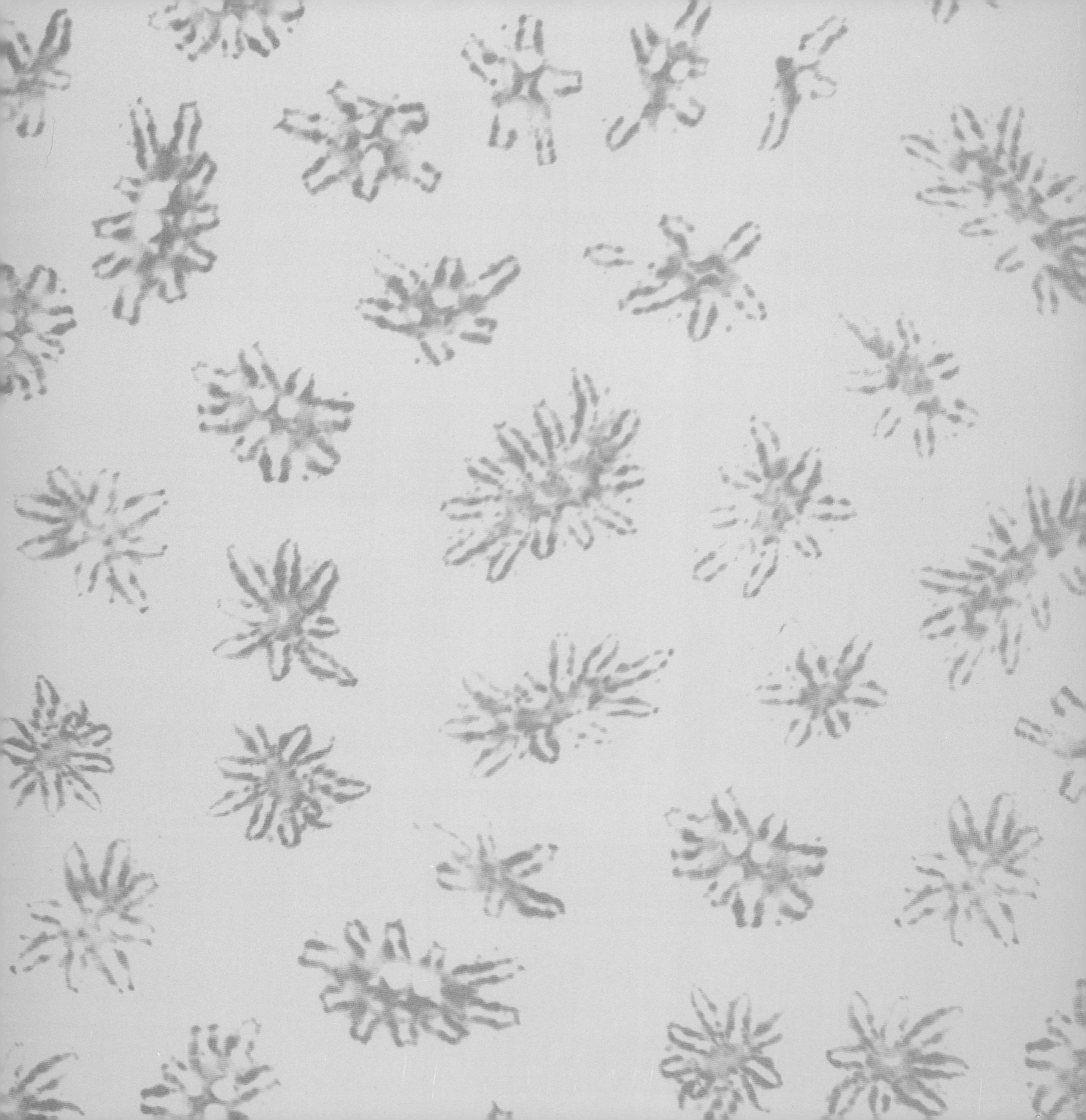

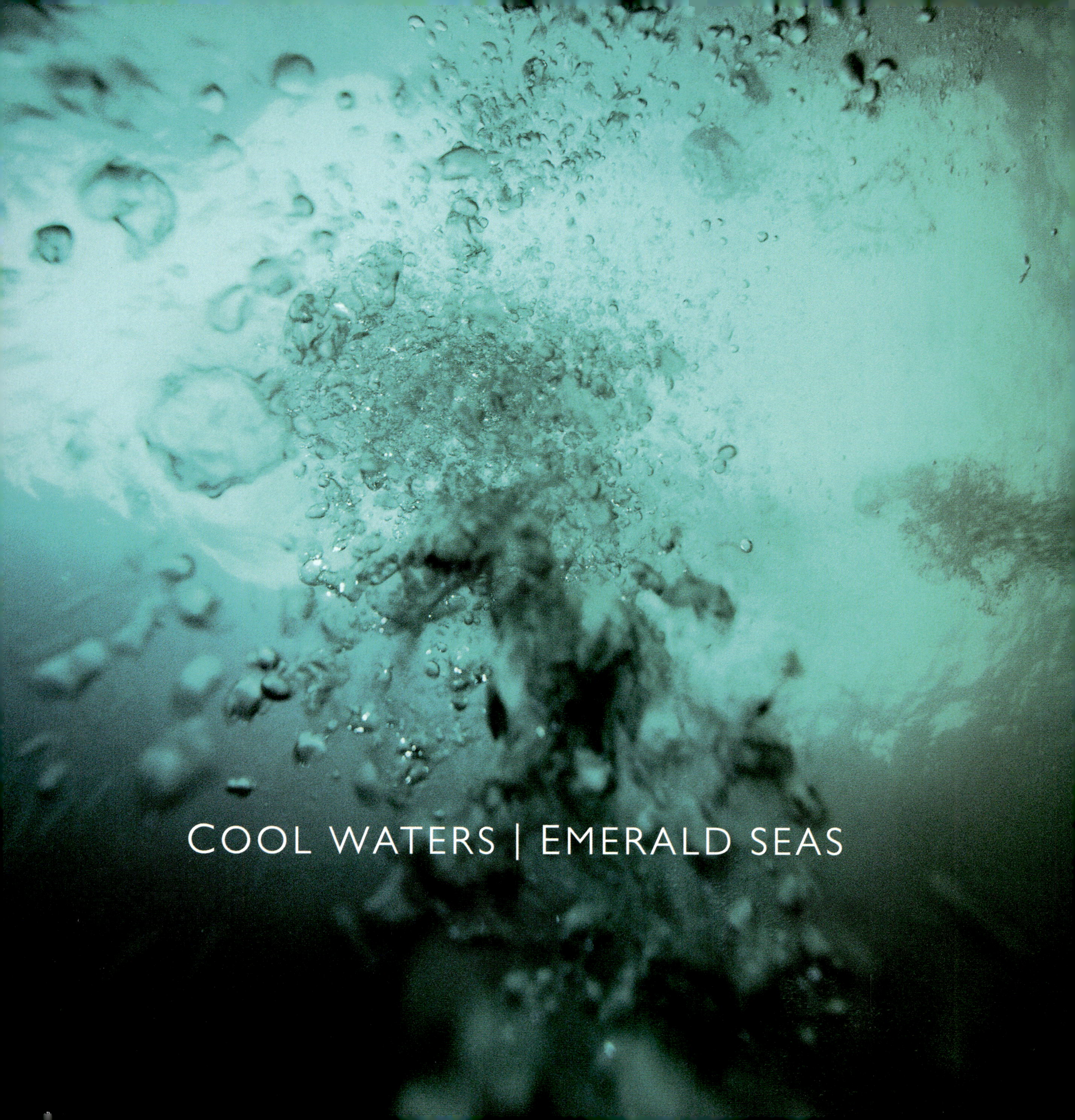

Cool Waters | Emerald Seas

ATRIUM

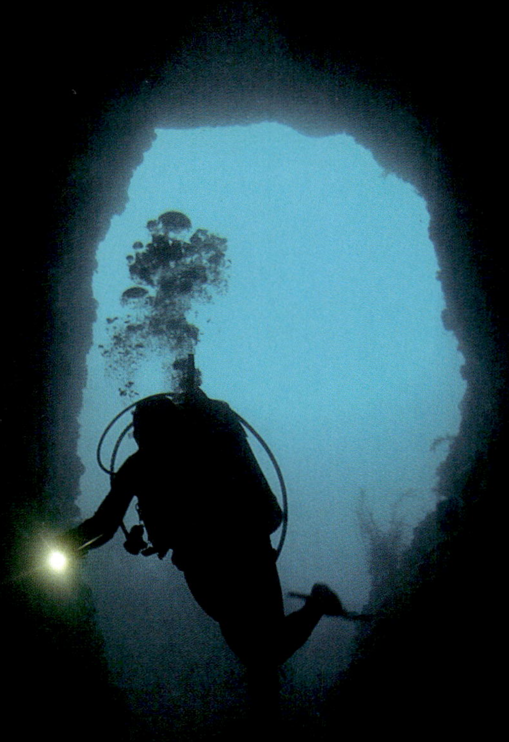

COOL WATERS | EMERALD SEAS

DIVING IN TEMPERATE WATERS

JOHN COLLINS

First published in 2006 by
Atrium
(an imprint of Cork University Press)
Youngline Industrial Estate
Pouladuff Road
Togher
Cork
Ireland

Designed by John Foley at Bite!, Cork
Printed in Ireland by ColourBooks Ltd, Baldoyle, Co Dublin

This publication has received support from the Heritage Council under the 2006 Publications Grant Scheme.

British Library Cataloguing in Publication data
A CIP catalogue record for this book is available from the British Library.

ISBN-10: 0-9535353-8-X
ISBN-13: 978-0-9535353-8-5
A CIP record for this publication is available from the Library of Congress

For all Atrium books visit www.corkuniversitypress.com

CONTENTS

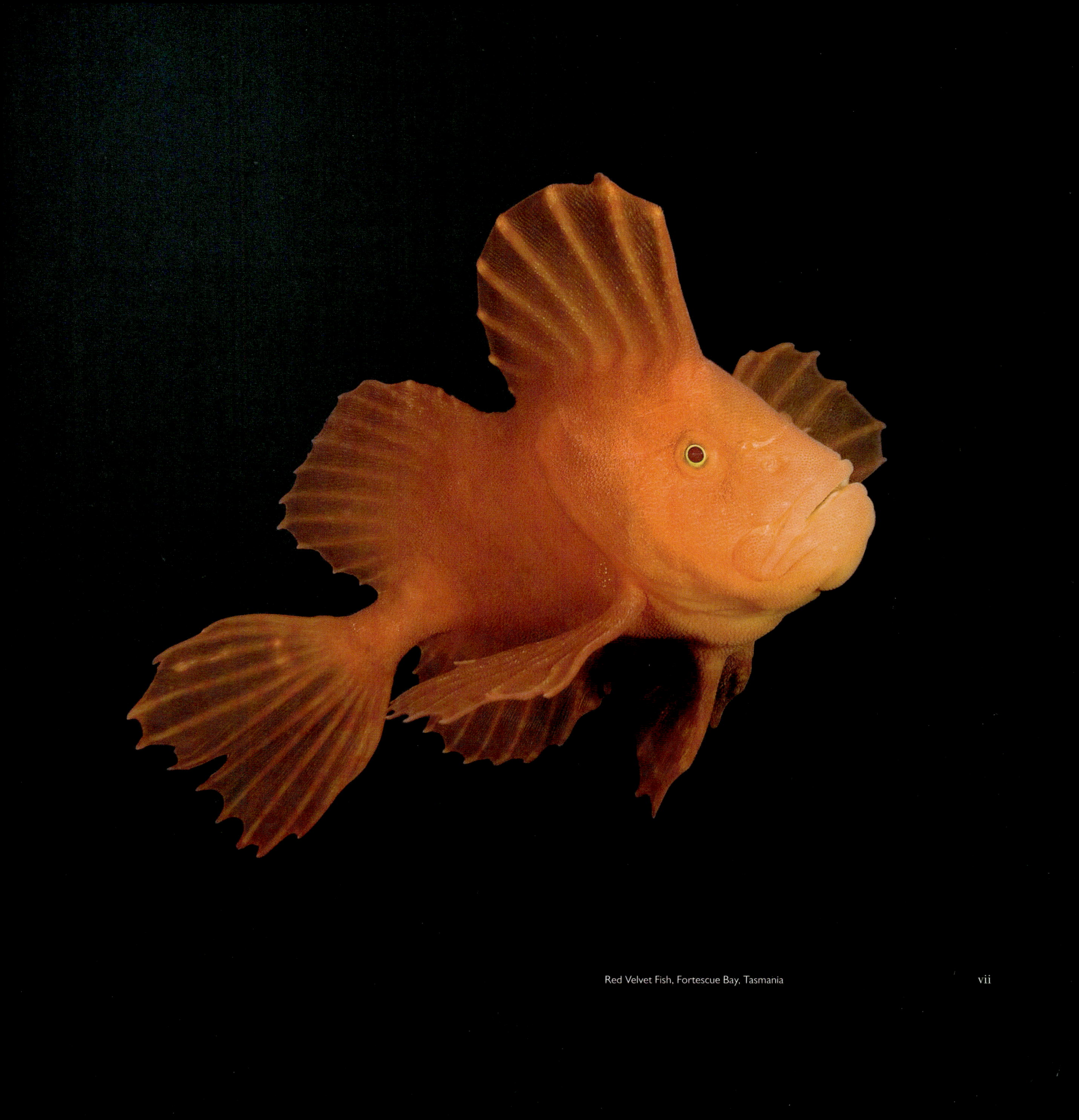

Red Velvet Fish, Fortescue Bay, Tasmania

ABOUT THE AUTHOR

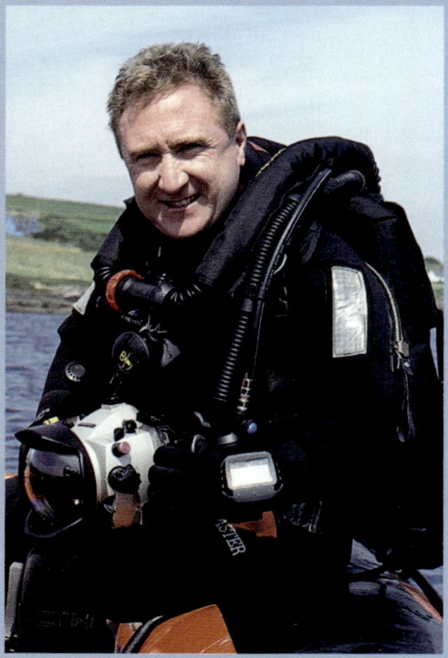

Photo: P. O'Sullivan

John Collins is an award-winning photographer based in Kinsale, Co. Cork. Originally from Limerick, he was educated at CBS Sexton Street and Trinity College Dublin. While at college he learned to scuba-dive with Dublin University Sub-Aqua Club and went on to gain experience both in Ireland and abroad, eventually becoming an instructor in 1991.

A passionate photographer from his teens, he was keen to photograph the underwater world and soon began to capture images of life beneath the waves. Many of these images have been successful in competition and widely published, both as written features and as stand alone photographs in magazines – including *Dive International*, *Sport Diver* (UK), *Diver Magazine*, *SubSea* and *Scuba World*.

John lives in Kinsale, on Ireland's south coast, with his wife Caoilfhionn and their two daughters Aislinn and Éadaoin, where they operate a family-run pharmacy.

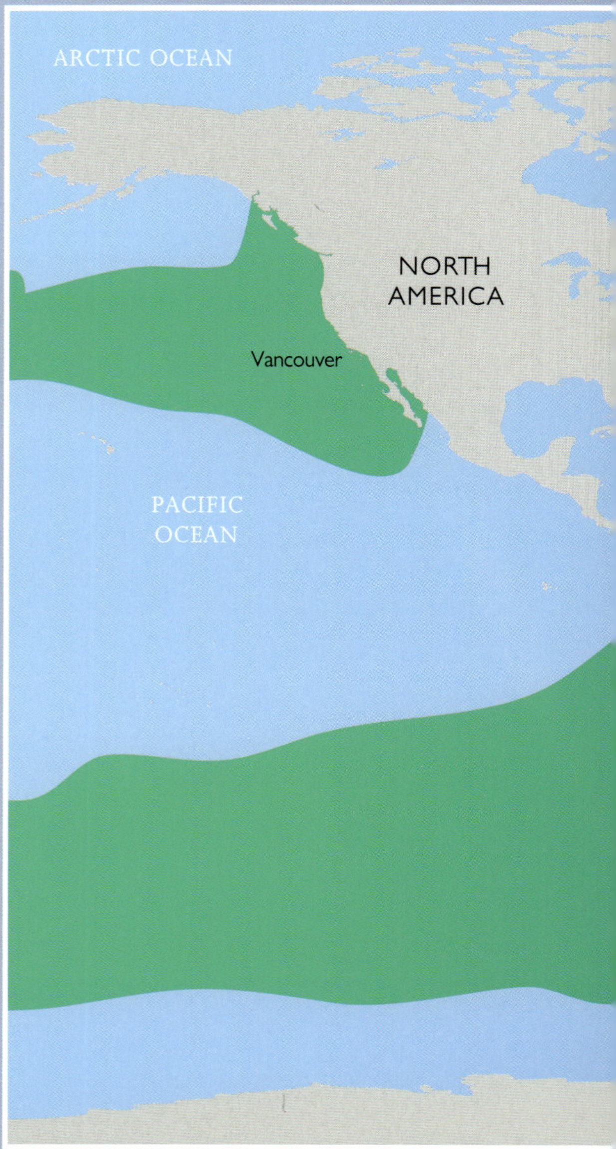

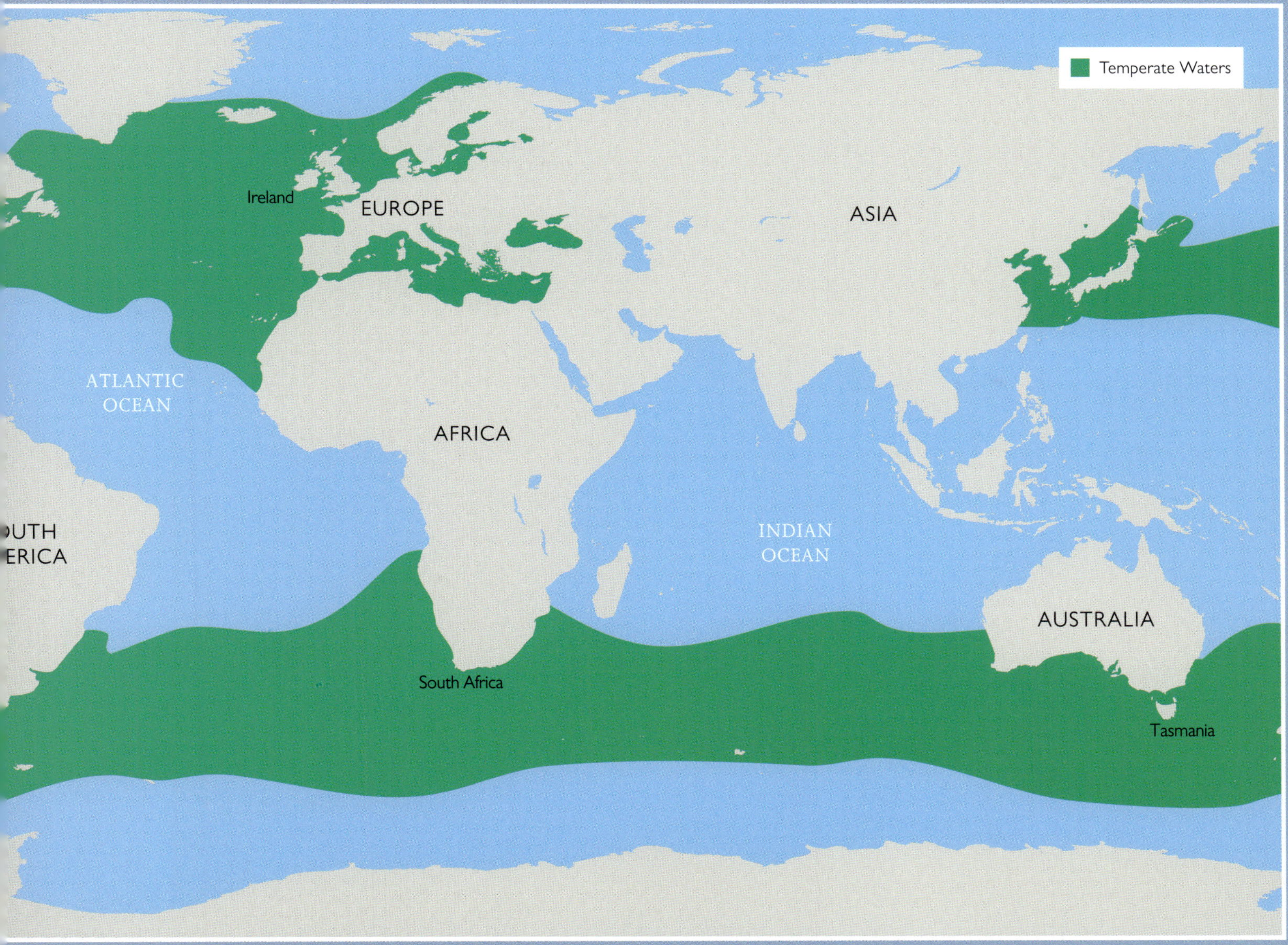
Temperate Waters
Ireland
EUROPE
ASIA
ATLANTIC OCEAN
AFRICA
INDIAN OCEAN
UTH
ERICA
AUSTRALIA
South Africa
Tasmania

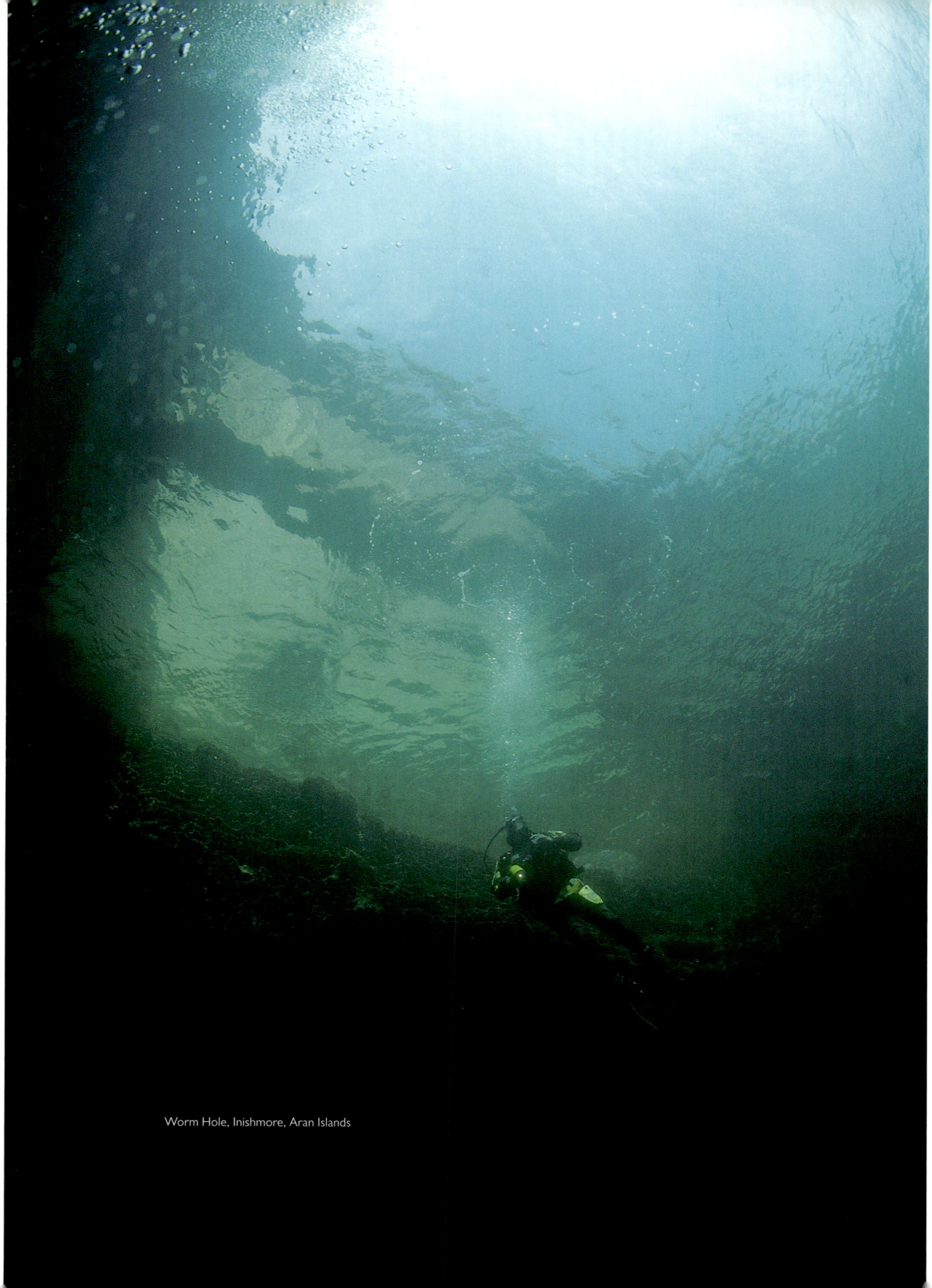

Worm Hole, Inishmore, Aran Islands

UNDERSEA JOURNEYS

I remember the first time I opened my eyes underwater. The blurred scene of a gravel river-bed gripped me; dappled light flickering across the plants and stones. A freshwater eel made a sudden dart from behind a rock, sending me bolting to the surface, gasping for breath. The noise of a summer's day returned, chaotic with children playing in the shallows. The dreamlike silence and mirage of that few seconds underwater was to begin a lifelong undersea journey.

The sea has a magnetism for the curious. It draws us as children to rummage in rock pools, hoping to see the simple antics of a crab, or to simply hold a starfish and wonder how it could be an animal at all. To see more, we donn a mask and snorkel and take to the water. The door to the undersea world opens a little and we glance inside, making short swims below, holding our breath. To explore further, we must learn to scuba dive.

To many, the idea of rolling over the side of a boat to swim to the seabed is quite unfathomable. The perception is that the sea is an alien place, not for us to visit because we cannot function underwater, as we can on land. To breathe we need cumbersome equipment and the senses that guide us in our air world are next to useless here. Vision requires a face-mask and is limited by the clarity of the water and, while hearing is acute, we cannot distinguish the source of sounds. It seems that in the millenia since we have evolved from the sea, we have completely lost the ability to venture underwater.

I often think how our predessors have struggled to explore the depths, mostly out of necessity to recover losses, or in wartime. Some of the attempts were ingenius and heroic, others foolhardy in the extreme. But, when Émile Gagnan and Jacques Cousteau perfected the aqua-lung in the 1940s, they gave us untethered freedom to explore the seas, and the new sport of scuba diving was born. Following in their footsteps, millions flock to tropical destinations every year to marvel at the magnificence of coral reefs. Yet, we overlook the waters nearer to us, perceiving them to be cold, barren and forbidding. But far from being dull and lifeless, our temperate seas are among the most vibrant on our water planet.

Learning to dive in the early 1980s, I realised that I was still among the early generations of Cousteau's 'lung-divers' to enjoy exploring the undersea world for the sheer pleasure and adventure it brings. As the Atlantic waters of Ireland slowly shared their secrets, I learned to capture the light and the life of the Emerald seas on camera. To see the variety of life in similar waters around the world, I journeyed to Vancouver island in the northwest Pacific, and to South Africa and Tasmania in the Southern Ocean. I have been continually amazed by the abundance and vibrance of ocean life in these seas.

Cool Waters, Emerald Seas is your armchair dive to the seabed of childhood imaginings. It is a portrait of the ocean's richest and most productive waters – the cool seas between the tropics and the polar regions. Unpredictable and seasonal, our explorations there are brief glimpses into a mesmerising world of life and colour. Sometimes it seems like the ocean is grooming itself for a visit, while we await fair weather. On a trip to Inishbofin, off Ireland's west coast, I waited for such an invitation for almost two weeks. While we were able to do some dives on the sheltered north shore, we knew that the waters to the west of the island would be special. With eleventh-hour drama, our opportunity to dive came on a calm, blue-skied final day. The sense of adventure of finally reaching the magnificent sea stack of Broughal rock remains with me, the underwater vista radiant in the afternoon light. The swell of the still active ocean rocked us gently as we dropped slowly down the wall, a kaleidoscope of colour all the way. While encounters with the big animals of the sea such as dolphins or sharks are the most exciting, the simple beauty of the static animals underwater are the most enchanting.

We have unwittingly made our own underwater vistas too. Chiselled coastlines, winter storms and wars have all taken their toll on shipping through the ages. Thousands of years of seafaring have left countless shipwrecks on the sea-bed. Diving these wrecks is especially rewarding. It grasps the senses and emotions, for not only do our eyes see the remains of once elegant ships but our souls empathize with those who sailed on them.

The emerald seas hold secrets that few of us get to see. Join me on a journey beneath the waves and you will see that this underwater world is captivating, wondrous and worthy of our protection.

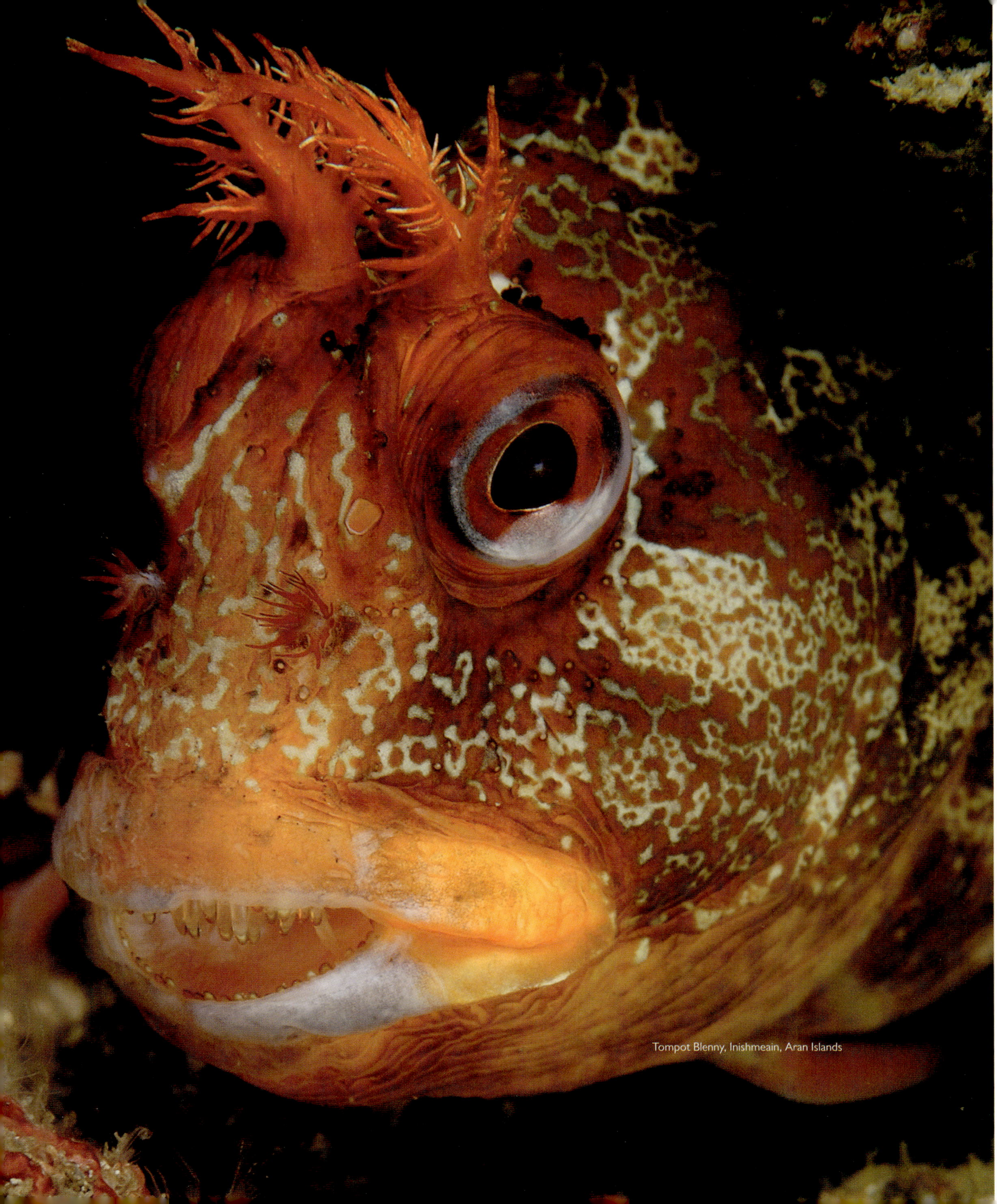

Tompot Blenny, Inishmeain, Aran Islands

One morning in June, 1943, I went to the railway station at Bandol on the French Riviera to collect a wooden case expressed from Paris. In it was a new and promising device, the result of years of struggle and dreams: an automatic compressed-air diving lung conceived by Émile Gagnan and myself. I rushed it to Villa Barry where my diving comrades, Philippe Tailliez and Frédéric Dumas, were waiting. No children ever opened a Christmas present with more excitement than we did when we unpacked the first 'aqualung'. If it worked, diving would be revolutionized.

We found an assembly of three moderate-sized cylinders of compressed air, linked to an air regulator the size of an alarm clock. The bottles contained air condensed to one hundred and fifty times atmospheric pressure. From the regulator there extended two tubes, joining on to a mouthpiece. With this equipment harnessed to the back, a water-tight glass mask over the eyes and nose, and rubber foot fins, we intended to make unencumbered flights in the depths of the sea.

Jacques Cousteau, *The Silent World*, 1953

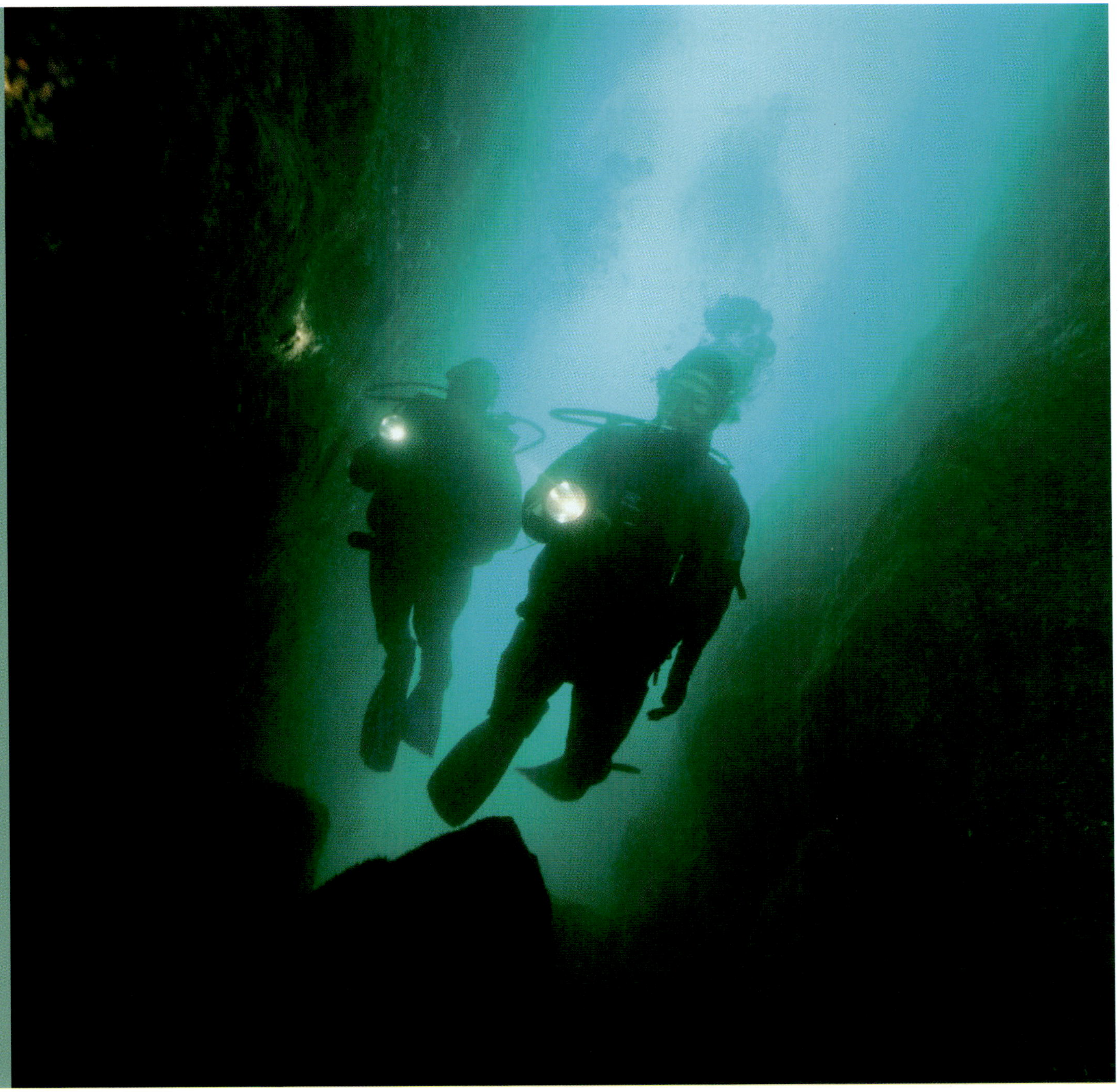

Carraig Coisceim, Inishturk Island, Co. Mayo

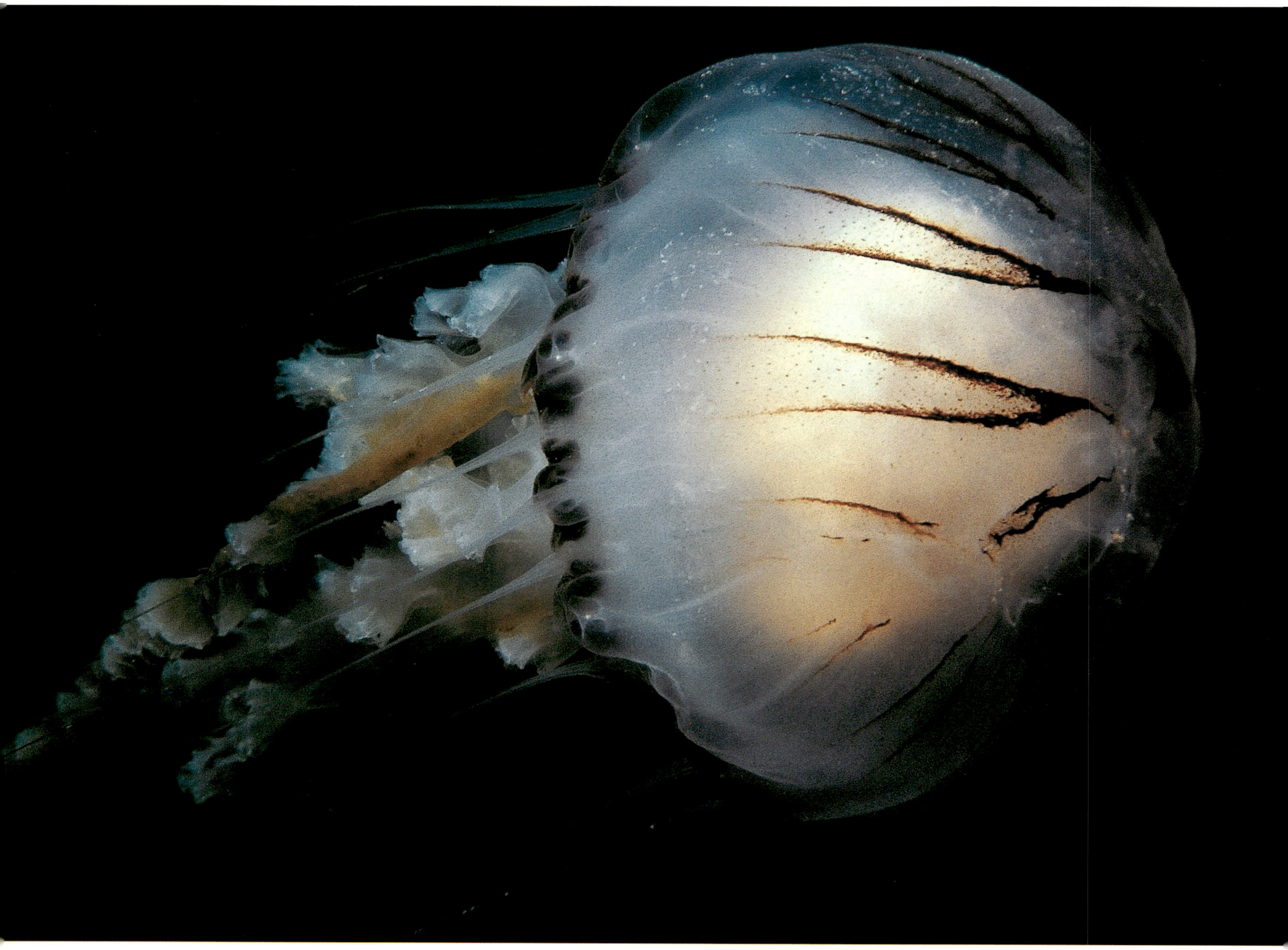

Compass Jellyfish, Keiss, Scotland

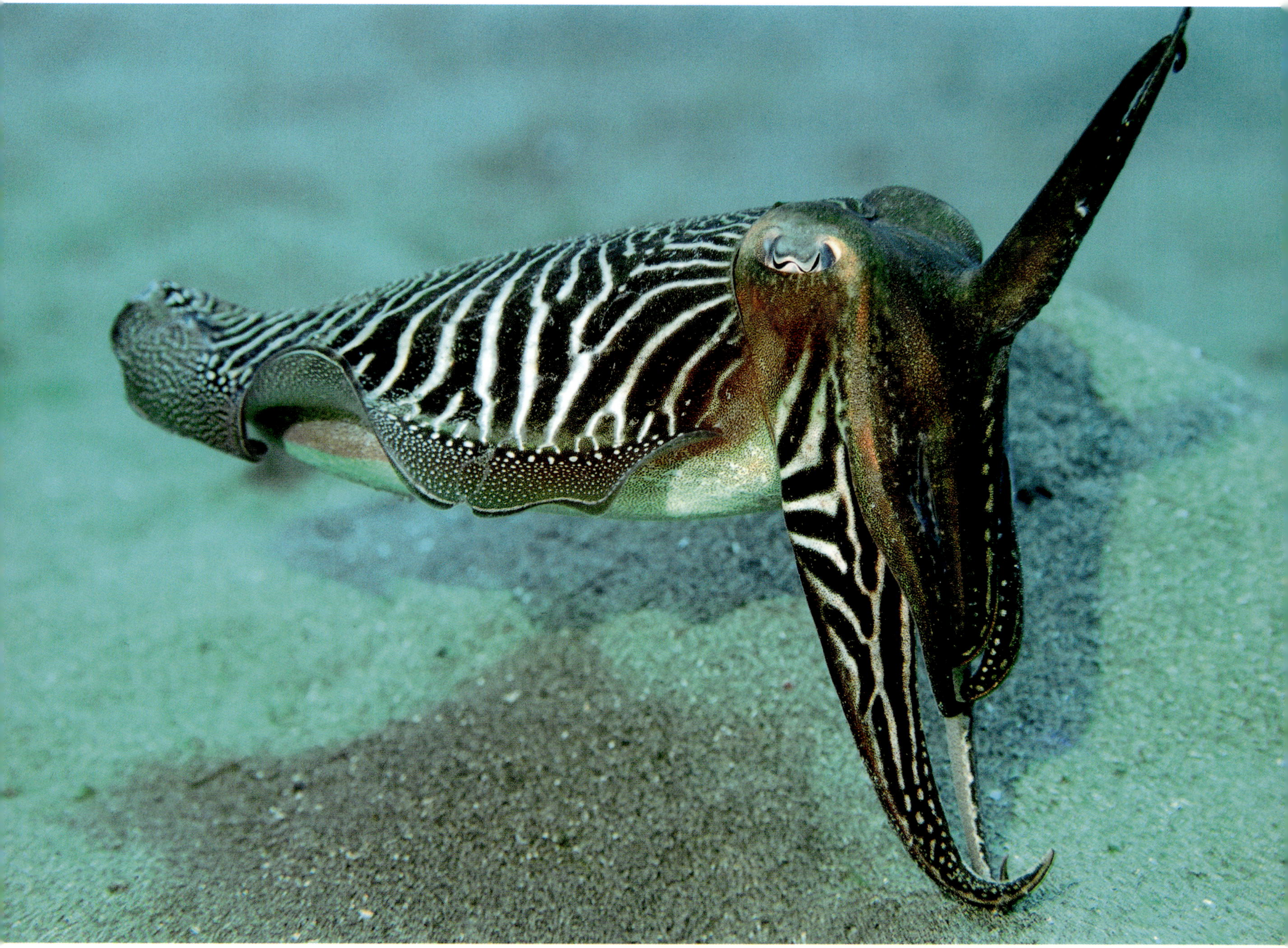

Cuttlefish, Lanzarote, Canary Islands

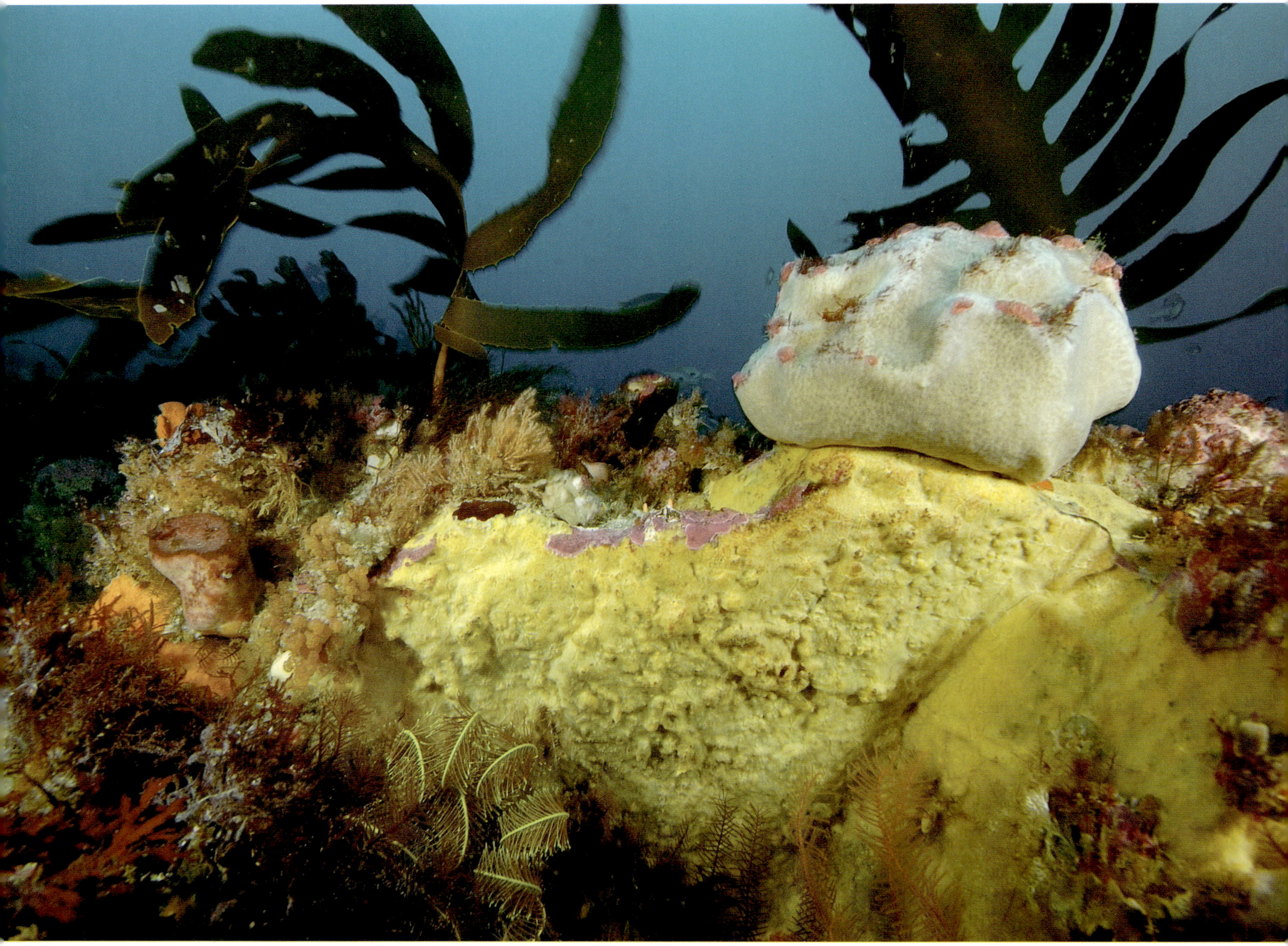

Sponge Garden, Sisters Rocks, Tasmania

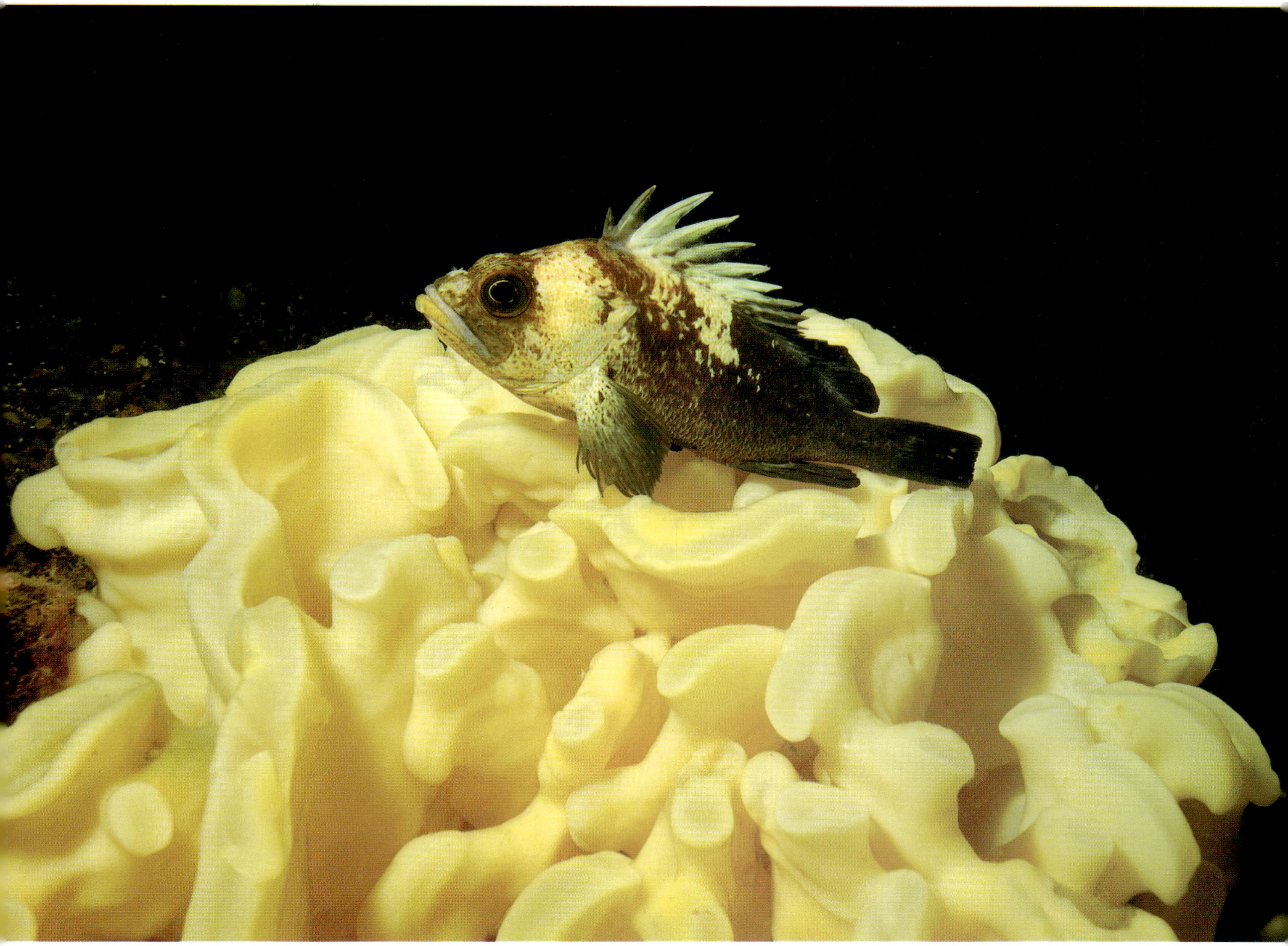

Quillback Rockfish on Cloud Sponge, Vancouver Island

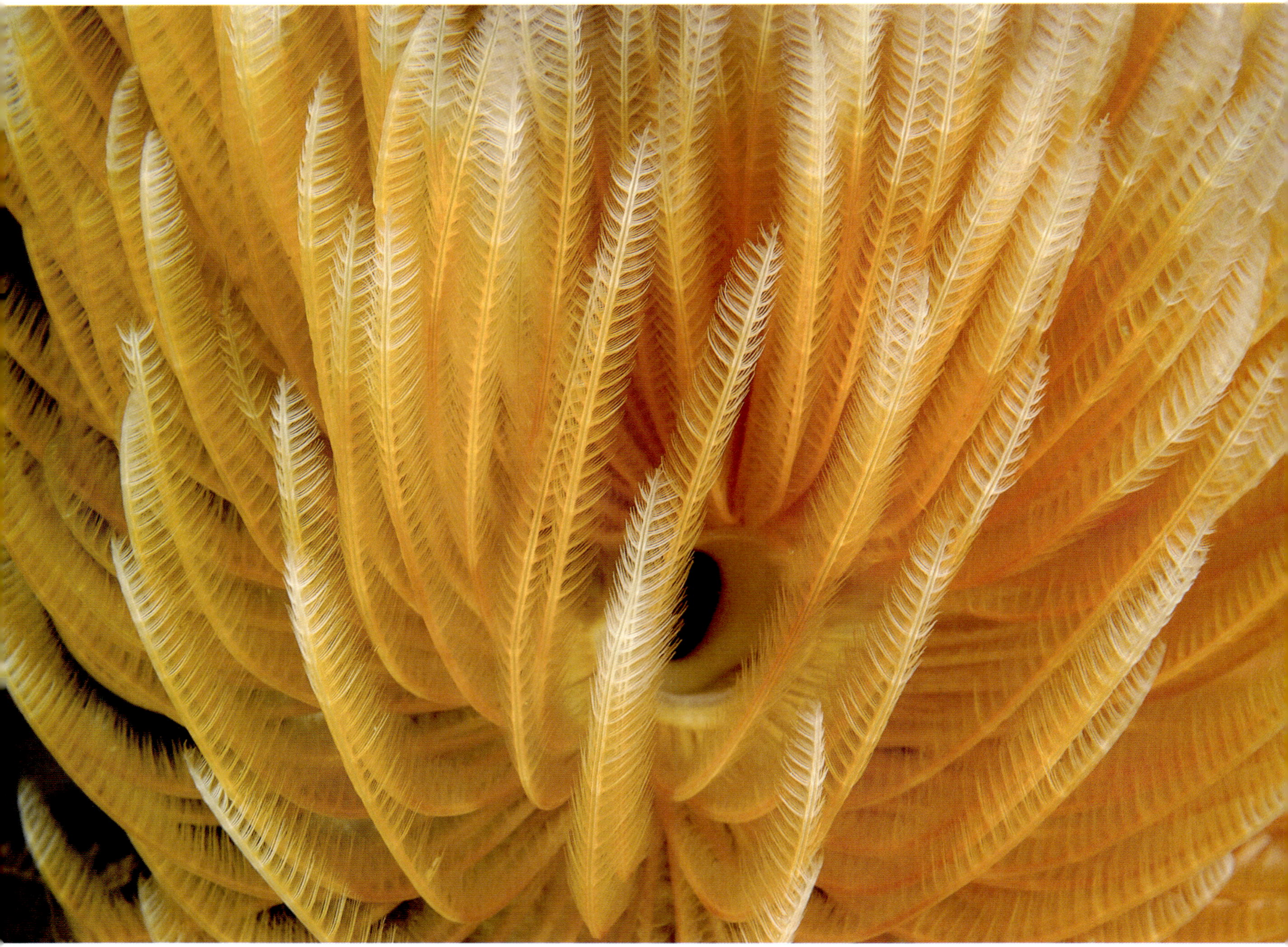

Feather Worm Detail, Waterfall Bay, Tasmania

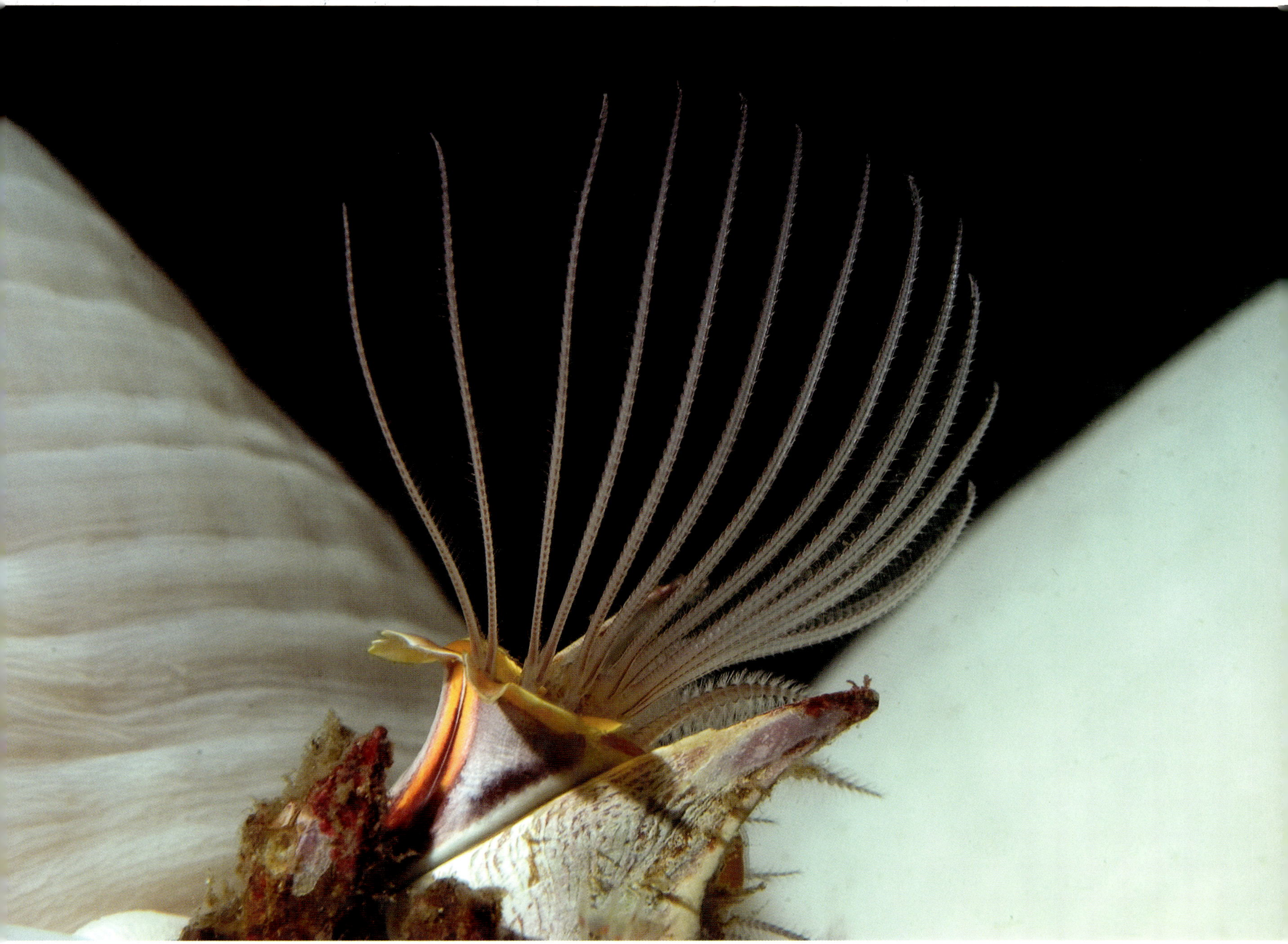

Goose Barnacle, Browning Pass, Vancouver Island

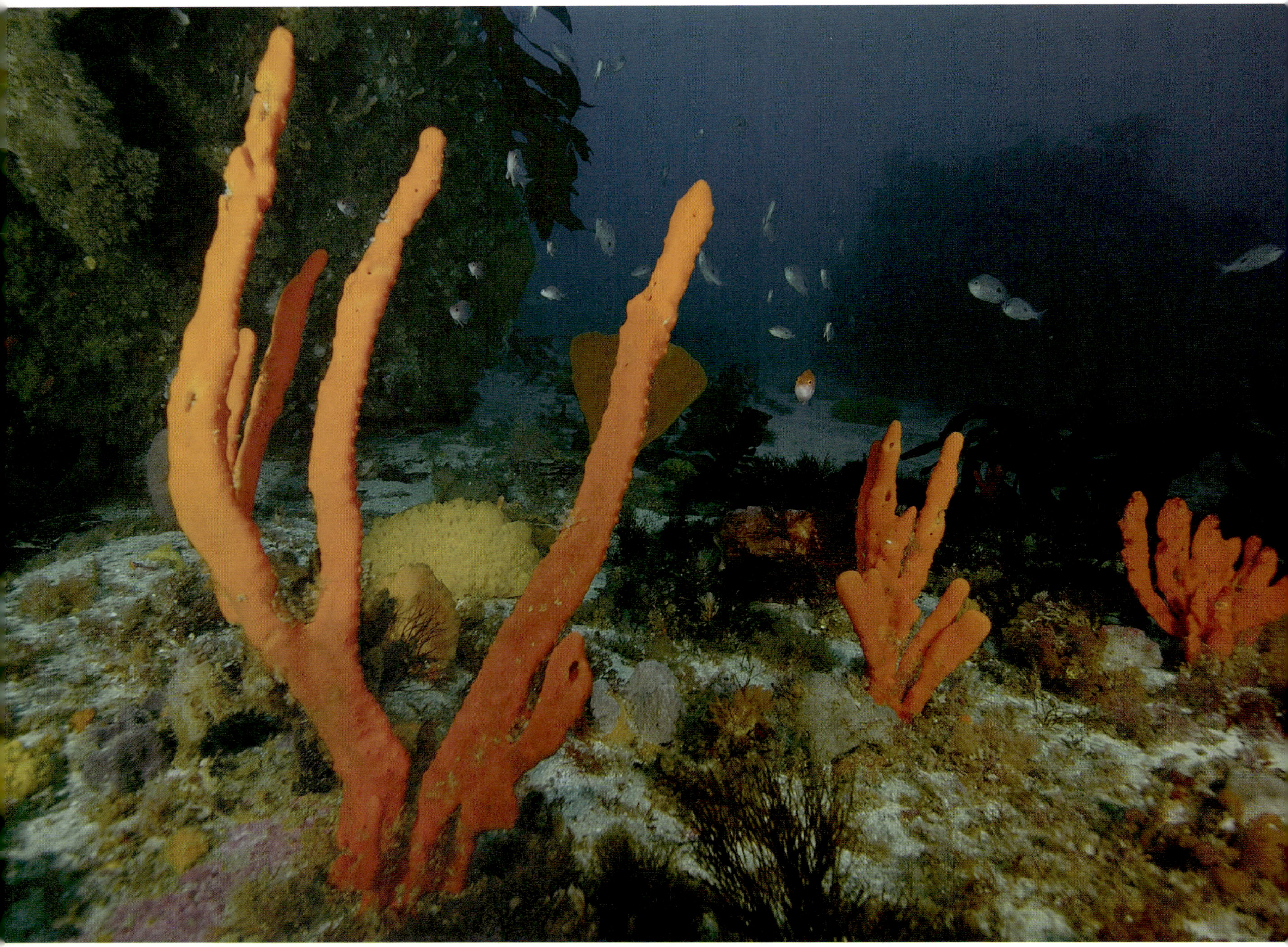

Sponge Garden, Sisters Rocks, Tasmania

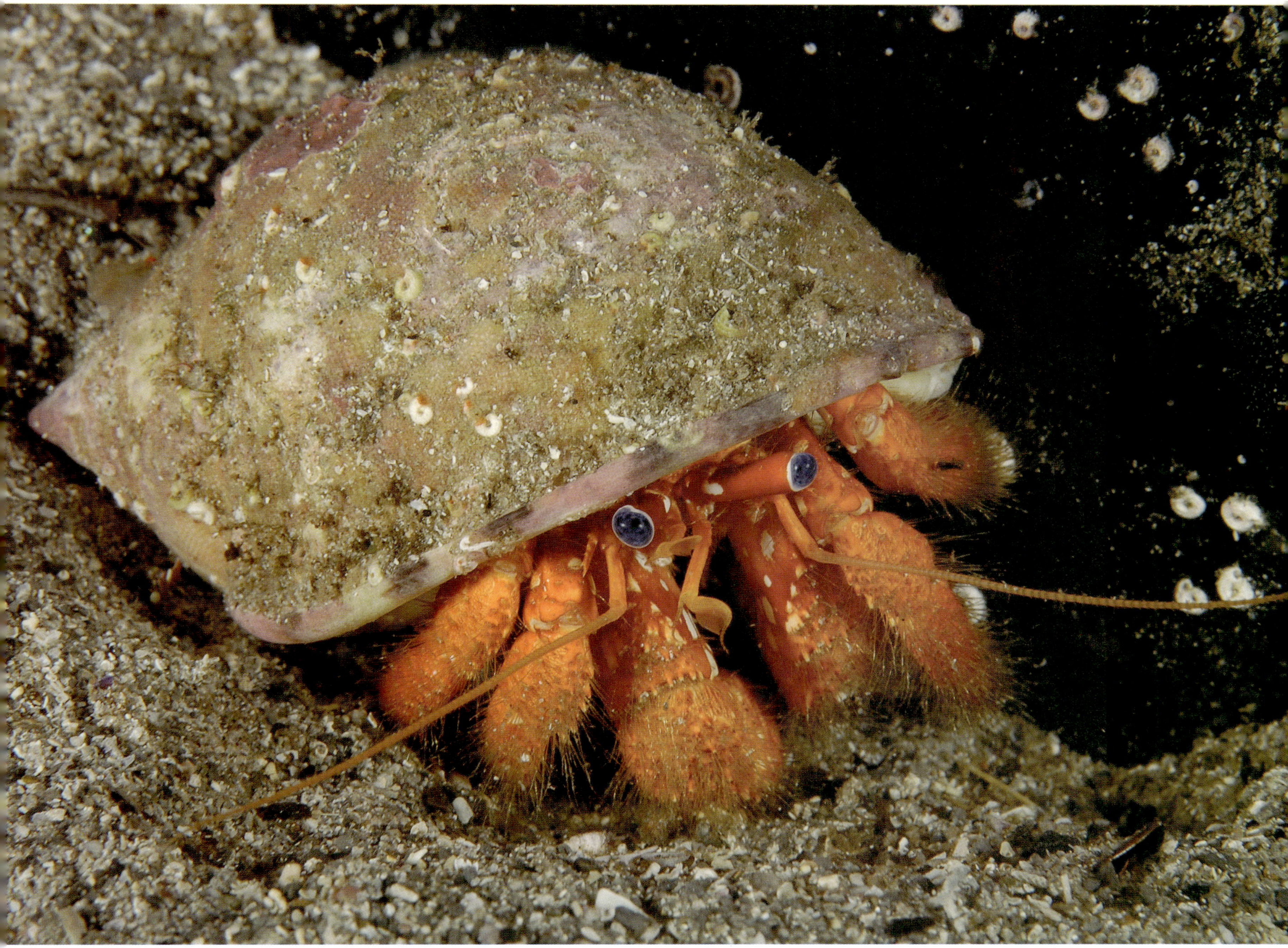

Hermit Crab, Pirates Bay, Tasmania

Each of us in turn tried out the self-operated diving gear. It kept its promises magnificiently. For the first time skin divers, who up to that time had to come up to the surface for lack of breath, experienced in a three-dimensional space, the free intoxication of diving without a cable. Back on the shore, we danced for joy. We were already certain that this was really the discovery which, a few years later, was to inaugurate, in France and the world over, the age of undersea exploration.

Captain Philippe Tailliez, *To Hidden Depths*, 1954

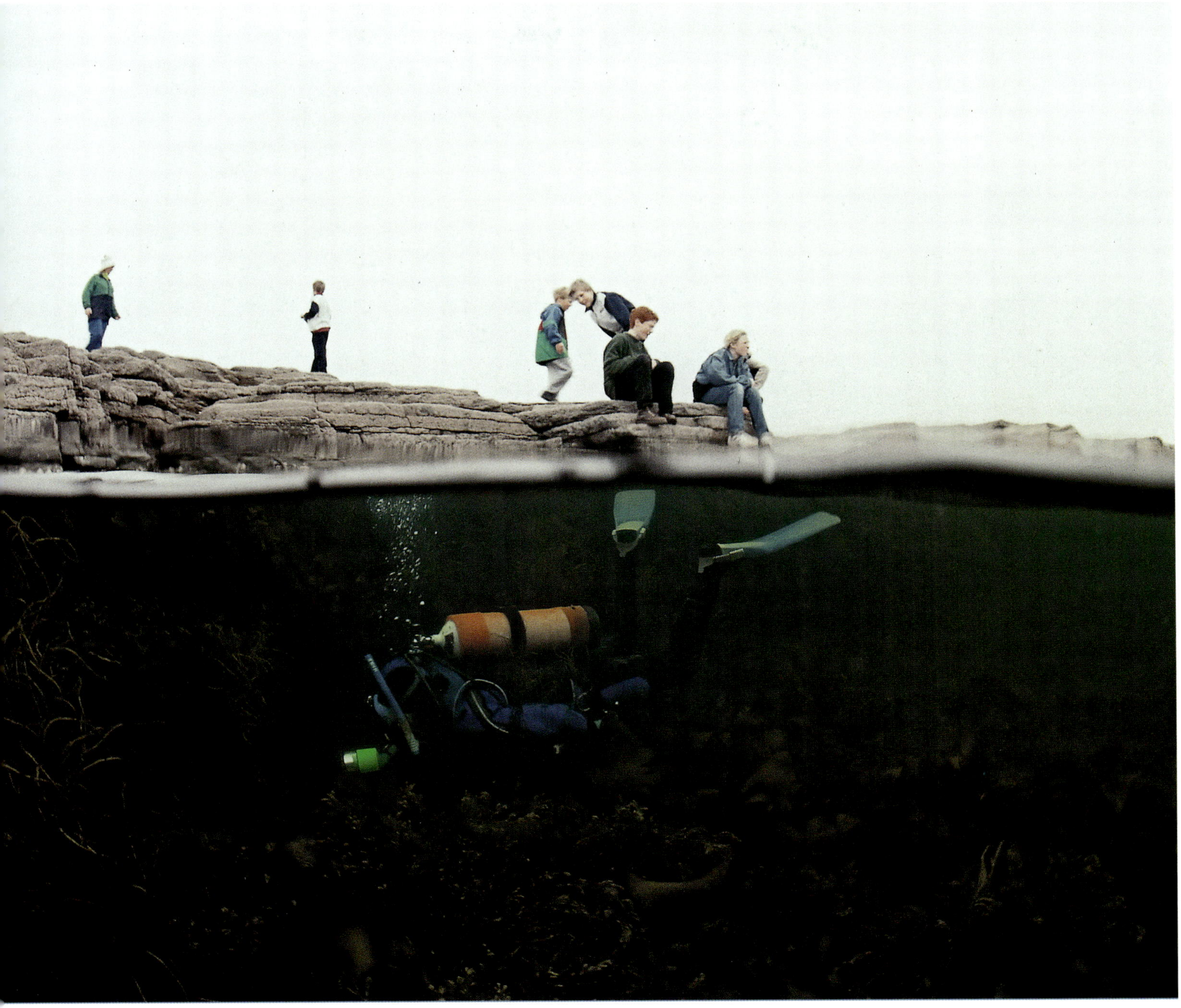

Pollack Holes, Kilkee, Co. Clare

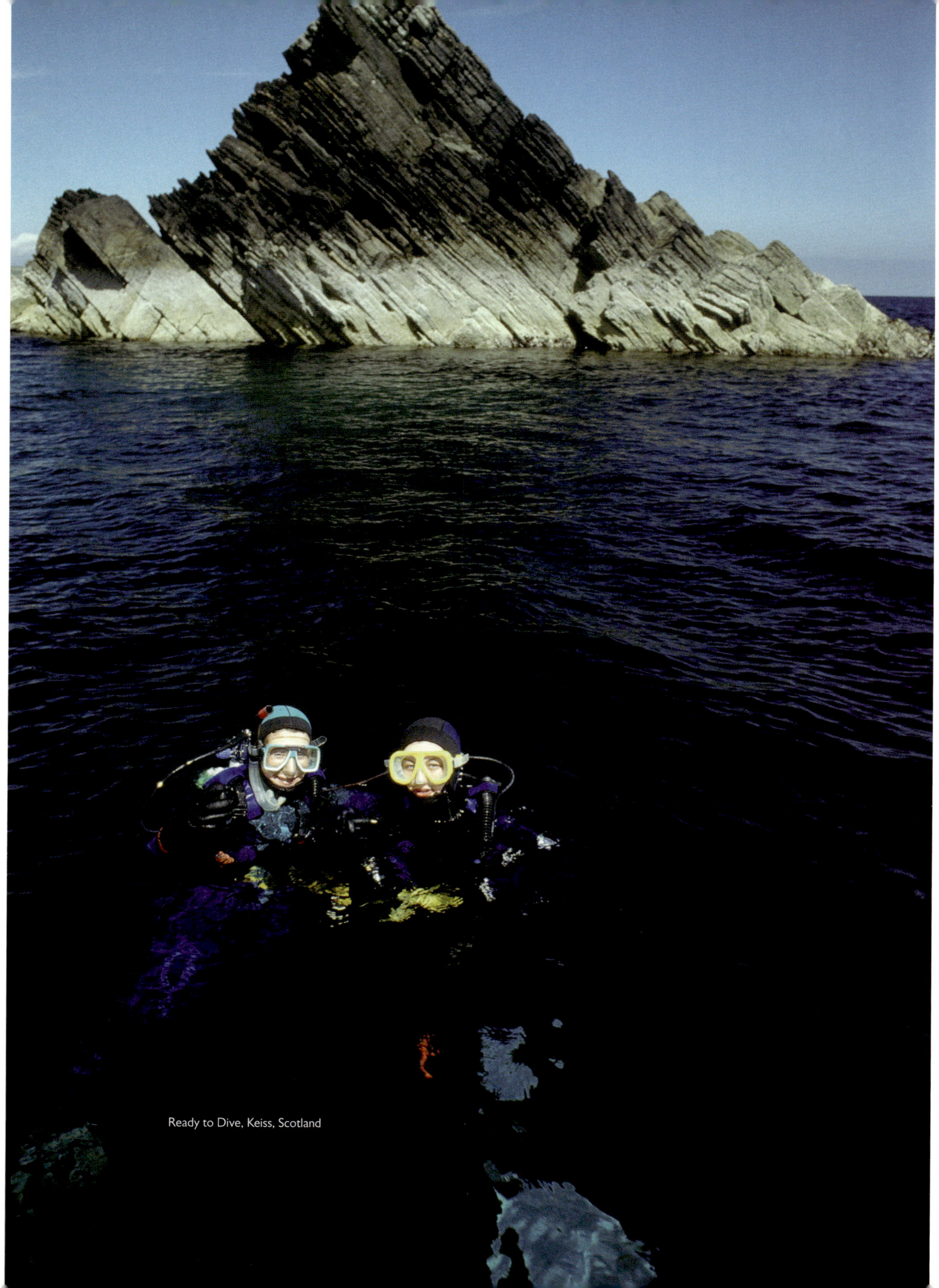

Ready to Dive, Keiss, Scotland

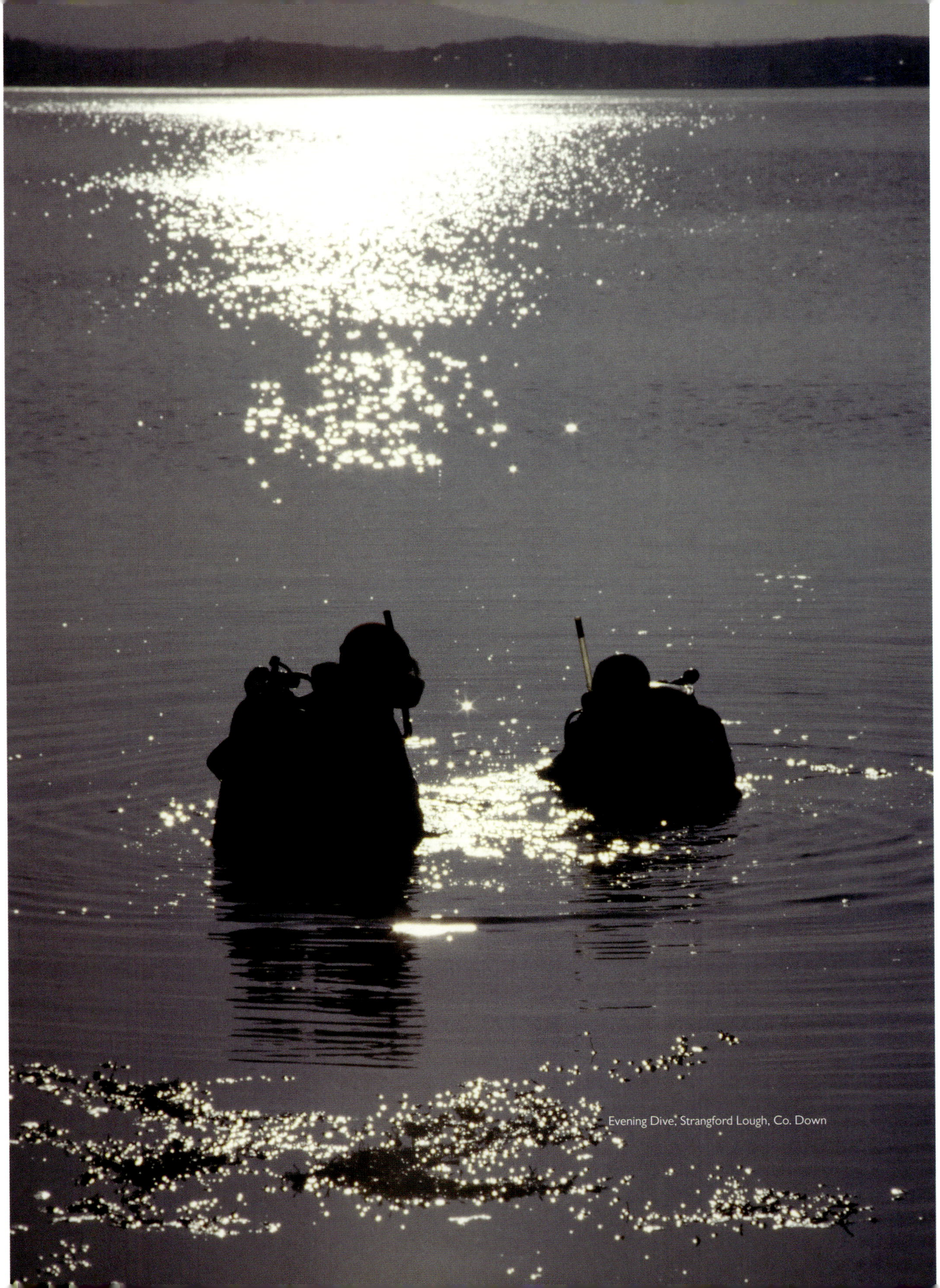

Evening Dive, Strangford Lough, Co. Down

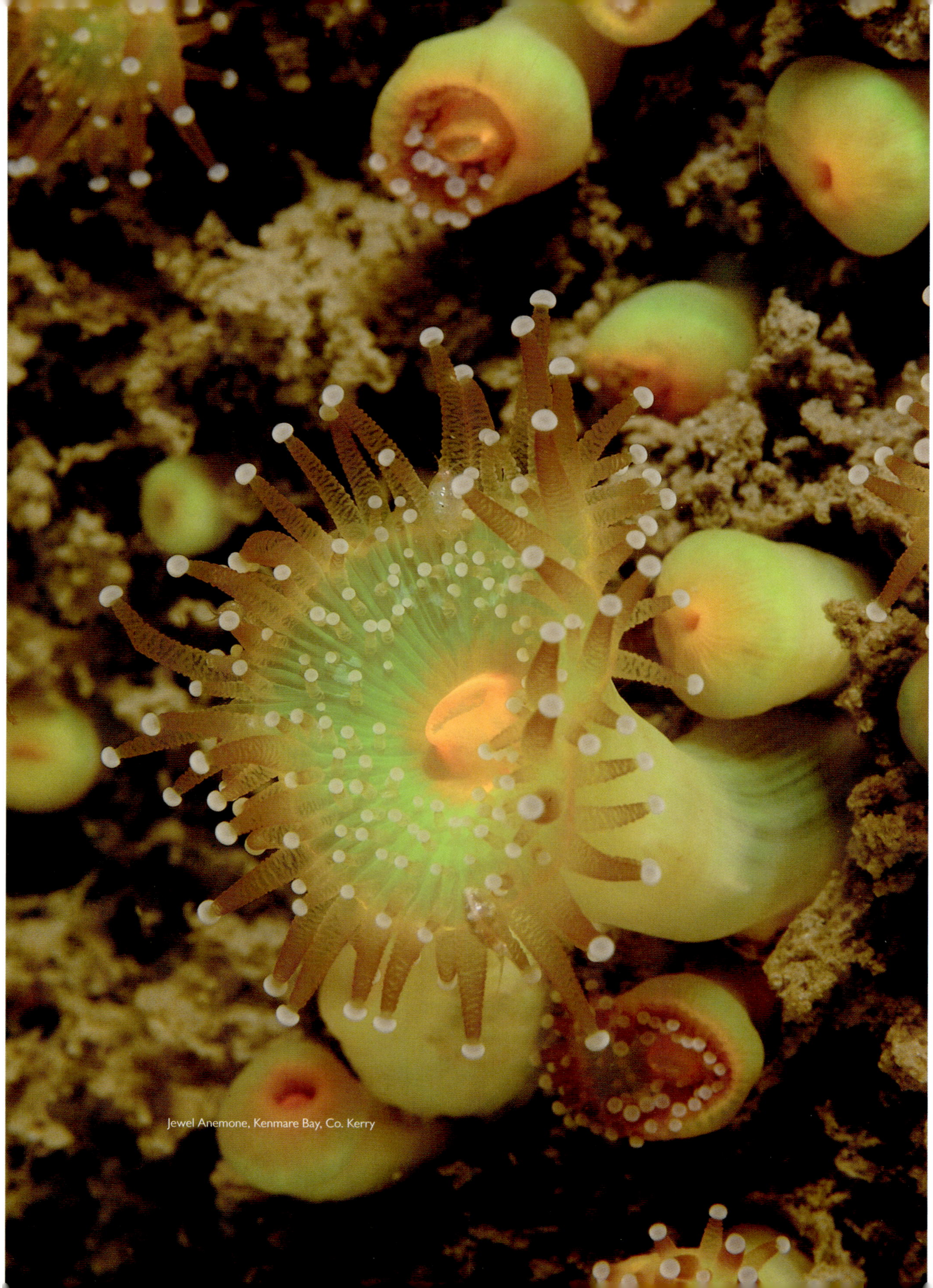

Jewel Anemone, Kenmare Bay, Co. Kerry

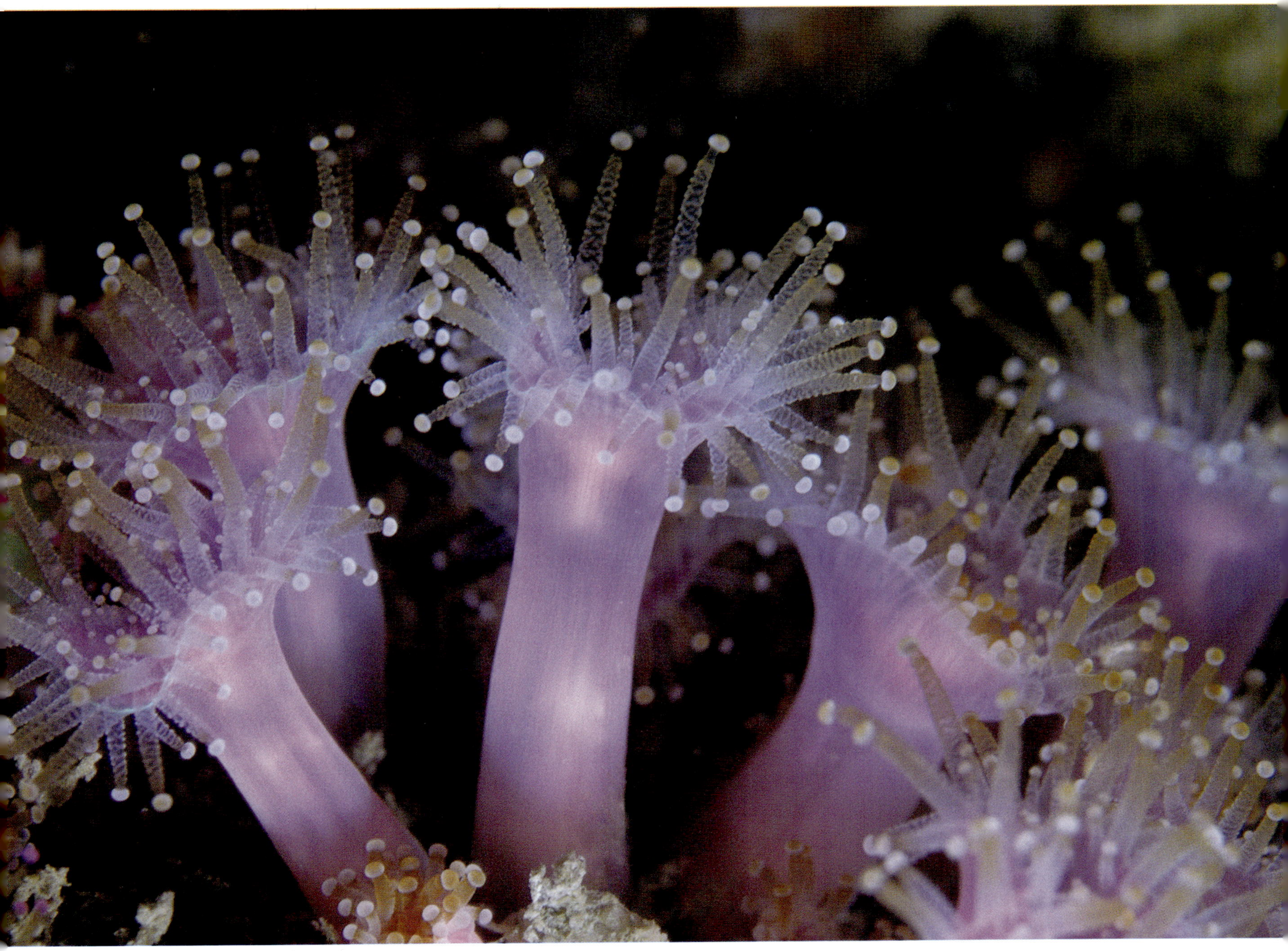

Jewel Anemones, St. John's Point, Co. Donegal

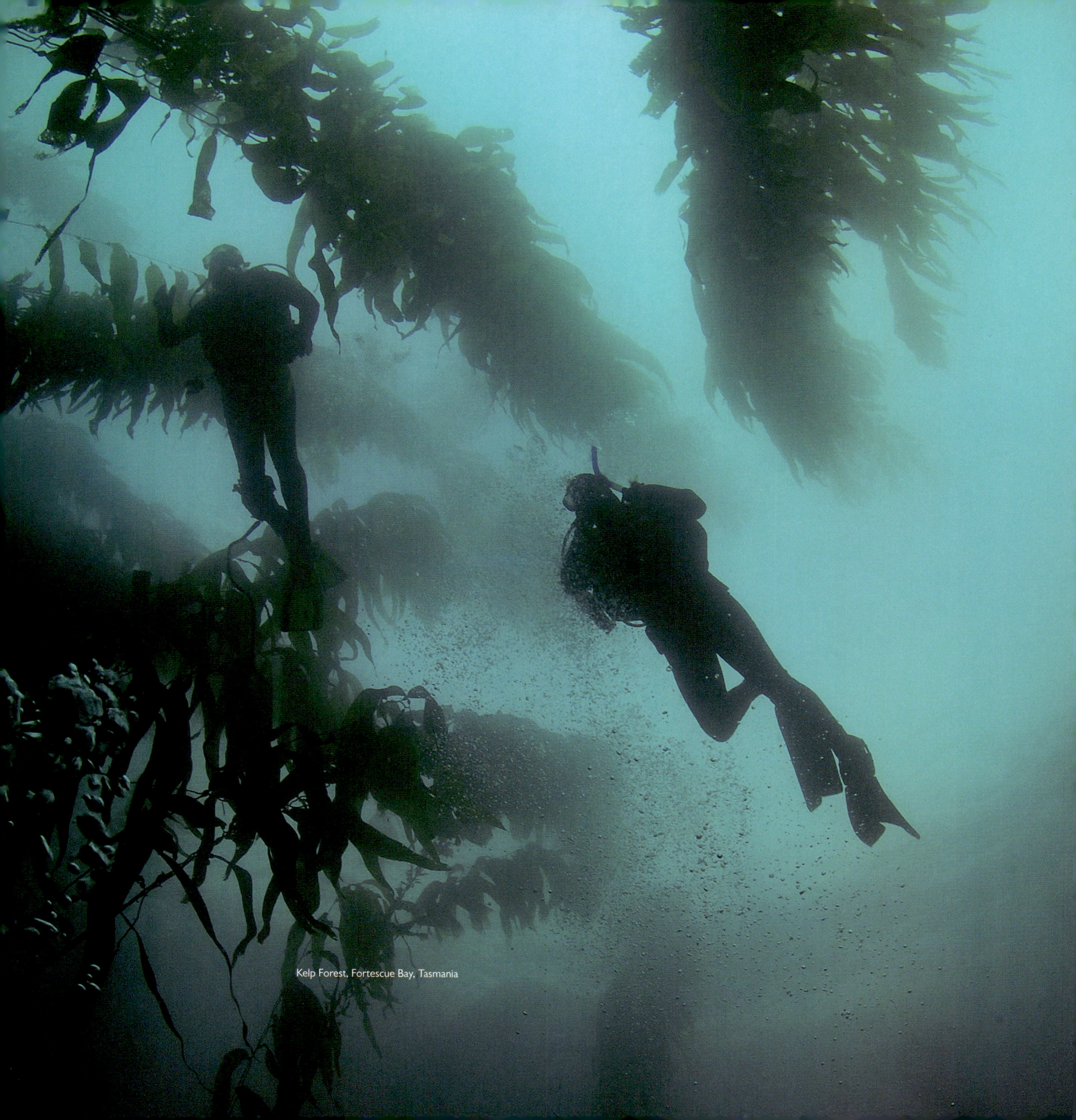

Kelp Forest, Fortescue Bay, Tasmania

EXPLORING COOL WATERS

A diver is most ungainly just before entering the water. Preparing for a dive beneath Ireland's Great Skellig, I am weighed down with equipment and uncomfortably warm at the surface. The skipper examines the echo sounder, which profiles the seabed, to find the ideal area – a rocky drop-off into deep water. Finally, the engine slows and he signals us to go. It is a relief to roll over the side of the boat and let the cool water take the weight. I release air from my buoyancy jacket and exhale to begin the dive. Once beneath the surface the transformation is complete – now weightless and graceful, I glide into a pale green wonderland.

This freedom to move so effortlessly in the sea, untethered to the surface, moved the French pioneers of scuba diving to write lyrical descriptions of how it felt to swim underwater. Cousteau wrote, 'I stood upside down on one finger and burst out laughing – a shrill distorted laugh. Whatever I did, nothing altered the automatic rhythm of air. Delivered from gravity and buoyancy, I flew around in space.' A contemporary of Cousteau's, Philippe Diolé found 'This underwater baptism, in the freshness of the spring sea, not only crowned me with riches from a kingdom that nobody could ever take away, but provided me with keys to unlock certain parts of myself.'

This revelation of three dimensional freedom greets every new diver. Learning to dive today remains a great personal adventure, more adequately experienced than described. The essential skills to use an aqua-lung safely are much the same as the early days, but the style of teaching has evolved. Modern dive courses focus on the new diver getting comfortable with equipment and learning personal limitations, as opposed to the militaristic survival training of years gone by. Most new divers are surprised to learn that our cool seas are the ocean's most productive. They are emerald green from the microscopic plants, or phytoplankton, that enrich them. Like land plants, these tiny organisms have the ability to turn sunlight into living tissue and so form the basis of all life in the oceans. They thrive on the nutrient and oxygen-rich water of the world's temperate seas and create the ocean's richest ecosystem.

The Skelligs are unique both above and below the surface. These two rocky islands, eight miles off the Kerry coast, are justly listed as being among the classic dives of the world. Leaving shore, they look small in the distance but as the tops of the rocks disappear into low cloud, their size is apparent. The morning sun lifts

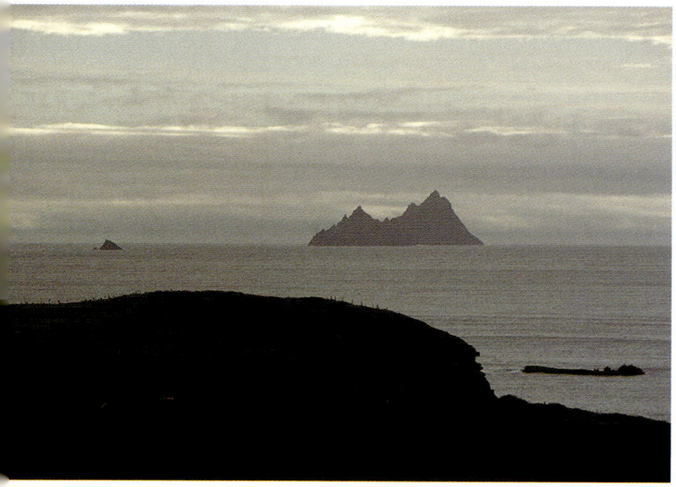

The Skelligs and Lemon Rock, from St. Finian's Bay

the cloud and the pointed tips are now visible, stretching skyward. Reaching over 200 metres high, the awe of their enormity is matched only by imagining what life was like for the monks who lived here over a thousand years ago. Beneath the craggy shoreline, the visiting diver is immersed within mother nature's cool waters and witness to a dazzling canvas in her portfolio.

By cool water we generally mean between four and twenty degrees Celsius. Worldwide, this is a huge part of our water planet. With seven tenths of the world's surface water rather than land, it is hard to grasp just how vast the oceans really are. We are told that from space our world looks as though it is predominantly water with continental patches of land. But because of the three-dimensional nature of water, the oceans contain 99.5 per cent of all the earth's living space. Most life occurs in coastal areas from the surface to around 200 metres depth – this is where sunlight reaches and where animals can attach, burrow or hide.

Gliding along Skellig's walls of encrusting life, the palette of colour could be that of a Cézanne or Gaugin painting. On this final dive of the week, I reflect on the preceding diving days. Memories of individual dives merge and the recollection of experiences seems like one long dive, akin to living within the sea for a few days. A curious cuckoo wrasse swims boldly to my face-mask, mystified by the strange looking intruder. It pauses, turns to one side, then another, as if checking with both eyes that it is not a mirage. Sensing no danger, it decides to join me as I swim along the wall. Orange soft corals blanket the rock face, their delicate white feeding polyps grasping at the passing water for a morsel of food. Every possible space is taken by large yellow sponges, gorgonians and colonies of multi-coloured anemones. Up close, tiny jewel anemones form a patchwork quilt of colour. Greens, yellows, neon pink and intense purple, glow in my torch-light. Starfish and urchins graze along the rocks and occasional gaps are taken by glowing light-bulb sea squirts.

Hunters abound too, nestled in the rocks during daylight. Large pie-shaped crabs back into their shelters, their huge claws giving them the appearance of flattened sumo wrestlers. An ageing, barnacle-encrusted lobster twiddles his antenna, curious about the strange invaders. He has escaped many seasons of lobster pots by making his migration to the depths for winter. Ascending gradually up the cliff face, I round the edge of the wall and spend the last few minutes of

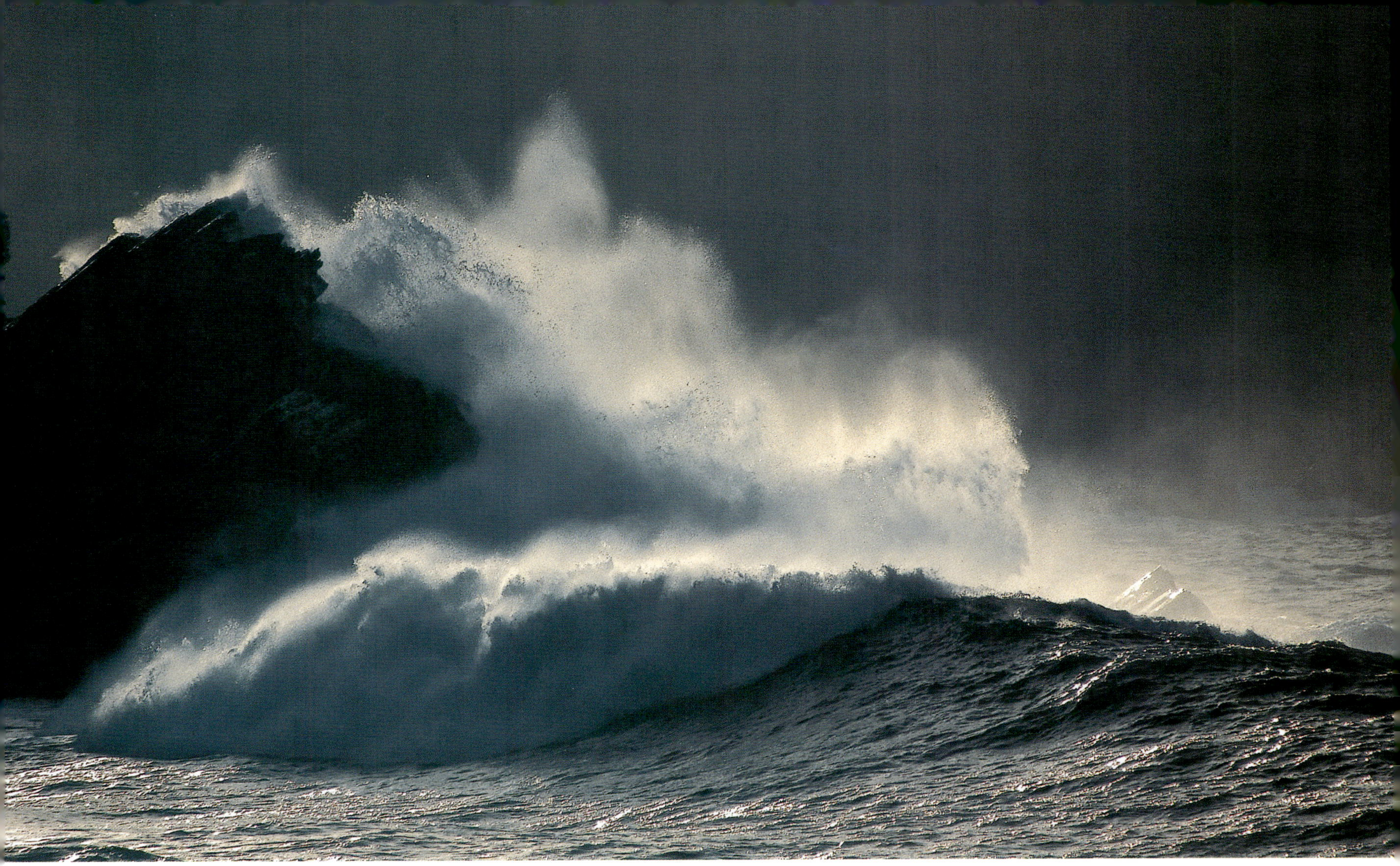

Waves, Ballyferriter, Dingle

the dive in the shallows. Here, among the plant life, a red scorpion fish has just ambushed a small pipe-fish and holds it in its mouth, wondering how to deal with such a substantial meal.

Approaching the surface, I glance around into open water and see a large compass jellyfish pulsating by. There are small movements among its tentacles. I swim closer to find a community of juvenile fish sheltering from predators, making short dashes out into open water to snatch a snack. The jellyfish's stinging tentacles, which are deadly to most passing prey have no effect on these little hitch-hikers, having cleverly developed a mucus layer to protect themselves.

Breaking the surface to see the boat and hear the birds once more, the spell has been broken and reality returns. To the west, an unblemished horizon reminds me that I have seen only a tiny part of the wondrous north Atlantic.

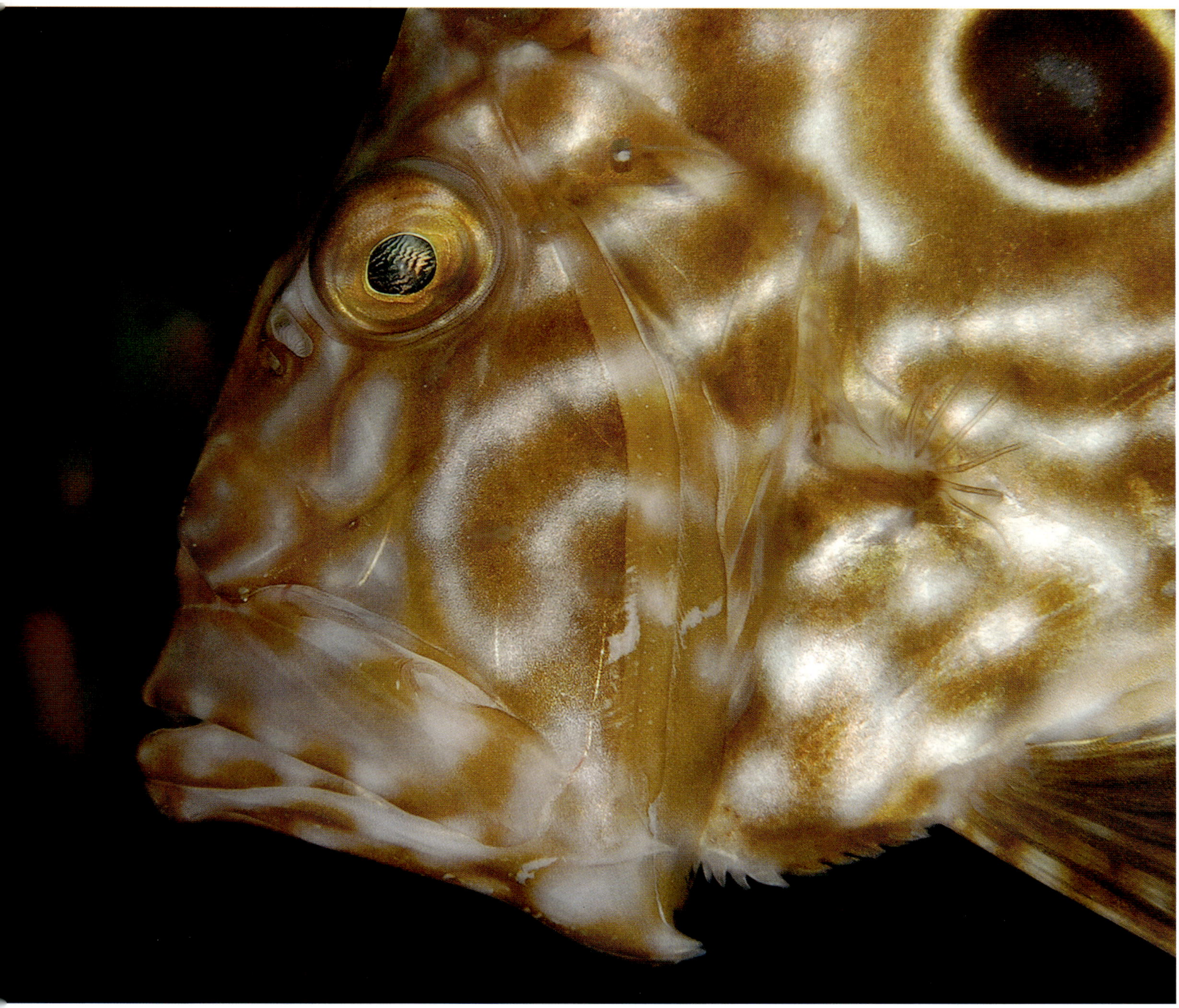

John Dory, Inisheer, Aran Islands

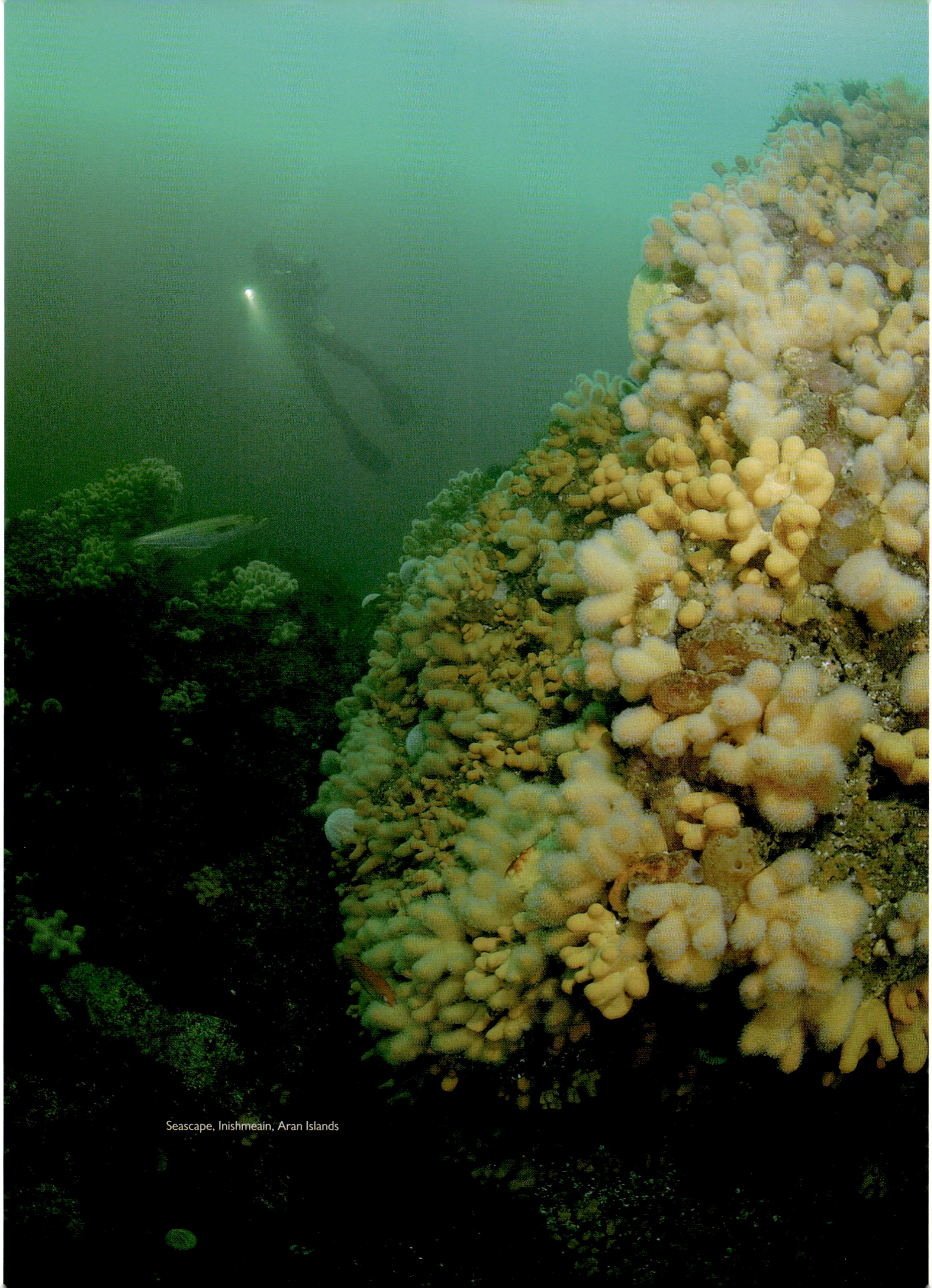

Seascape, Inishmeain, Aran Islands

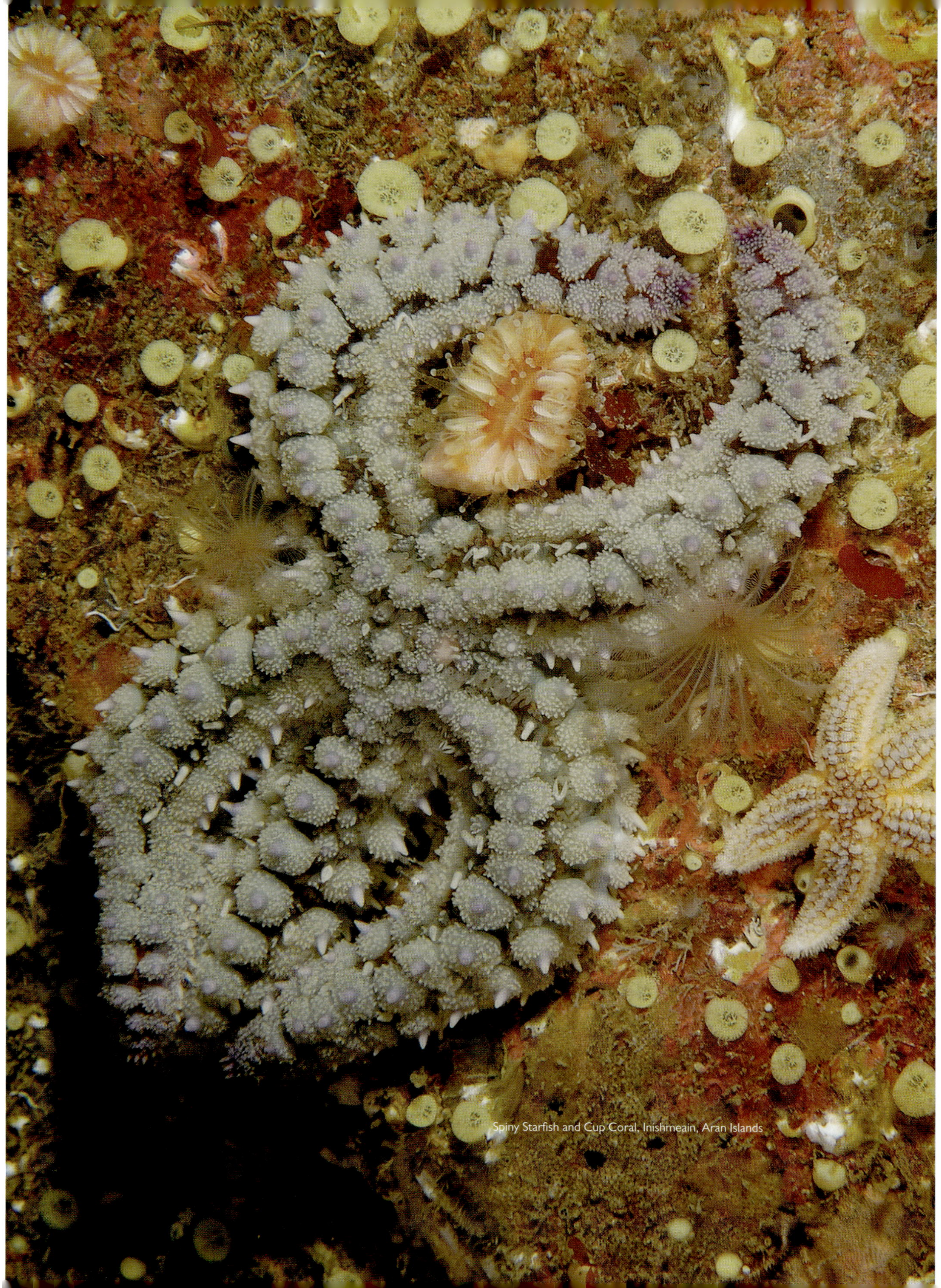

Spiny Starfish and Cup Coral, Inishmeain, Aran Islands

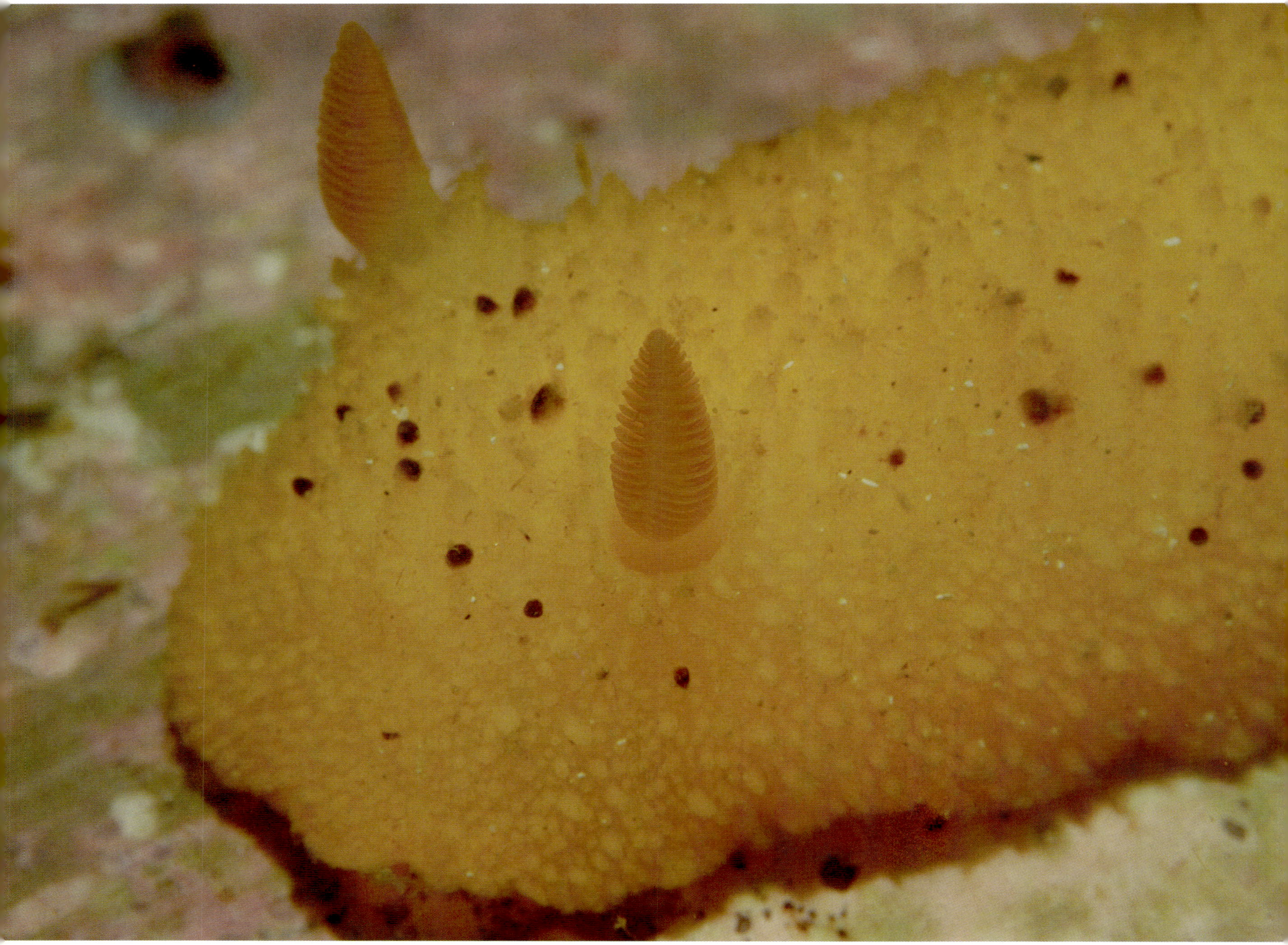

Sea Lemon Nudibranch, Hayden Pass, Vancouver

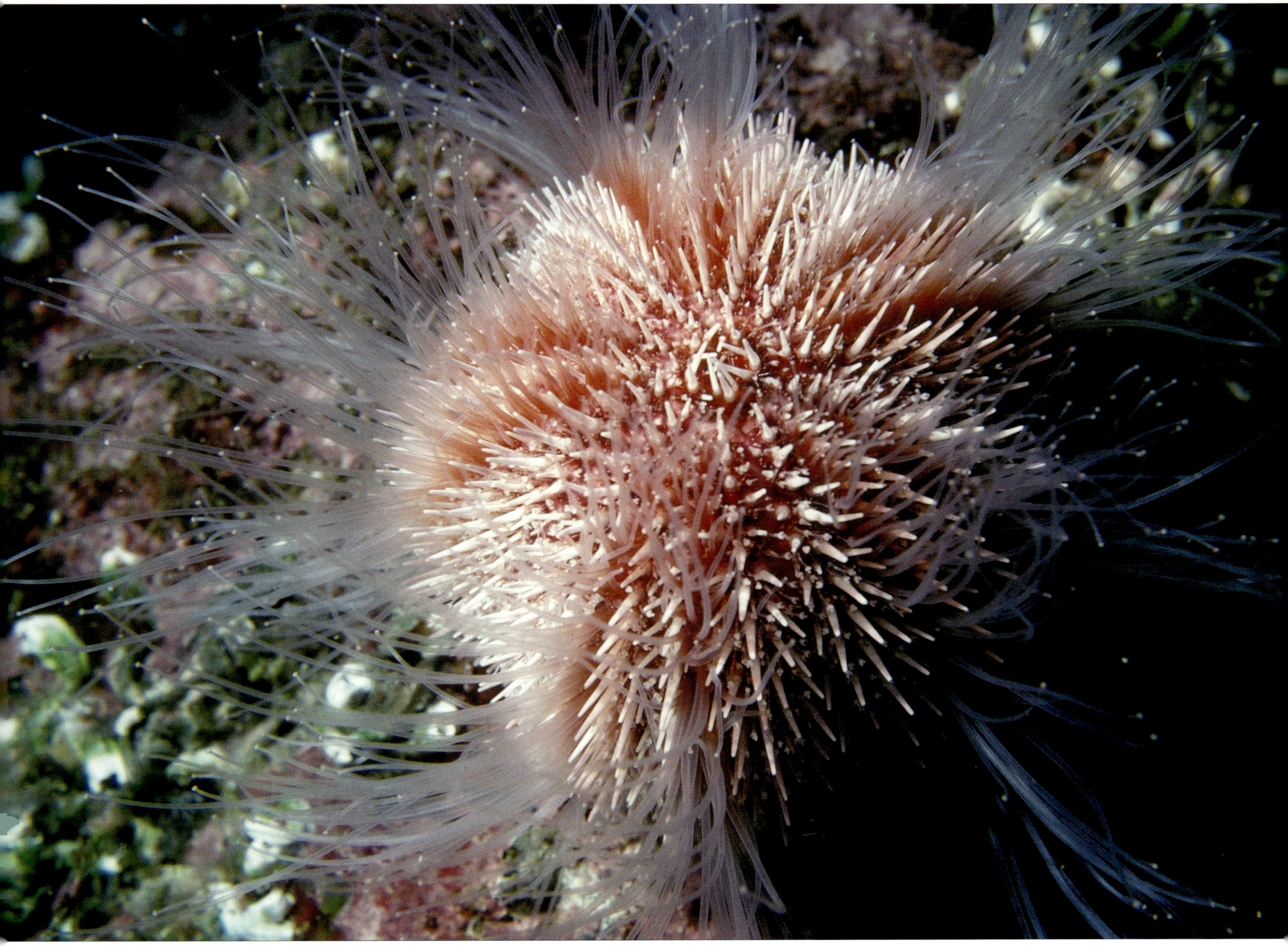

Sea Urchin, Keiss, Scotland

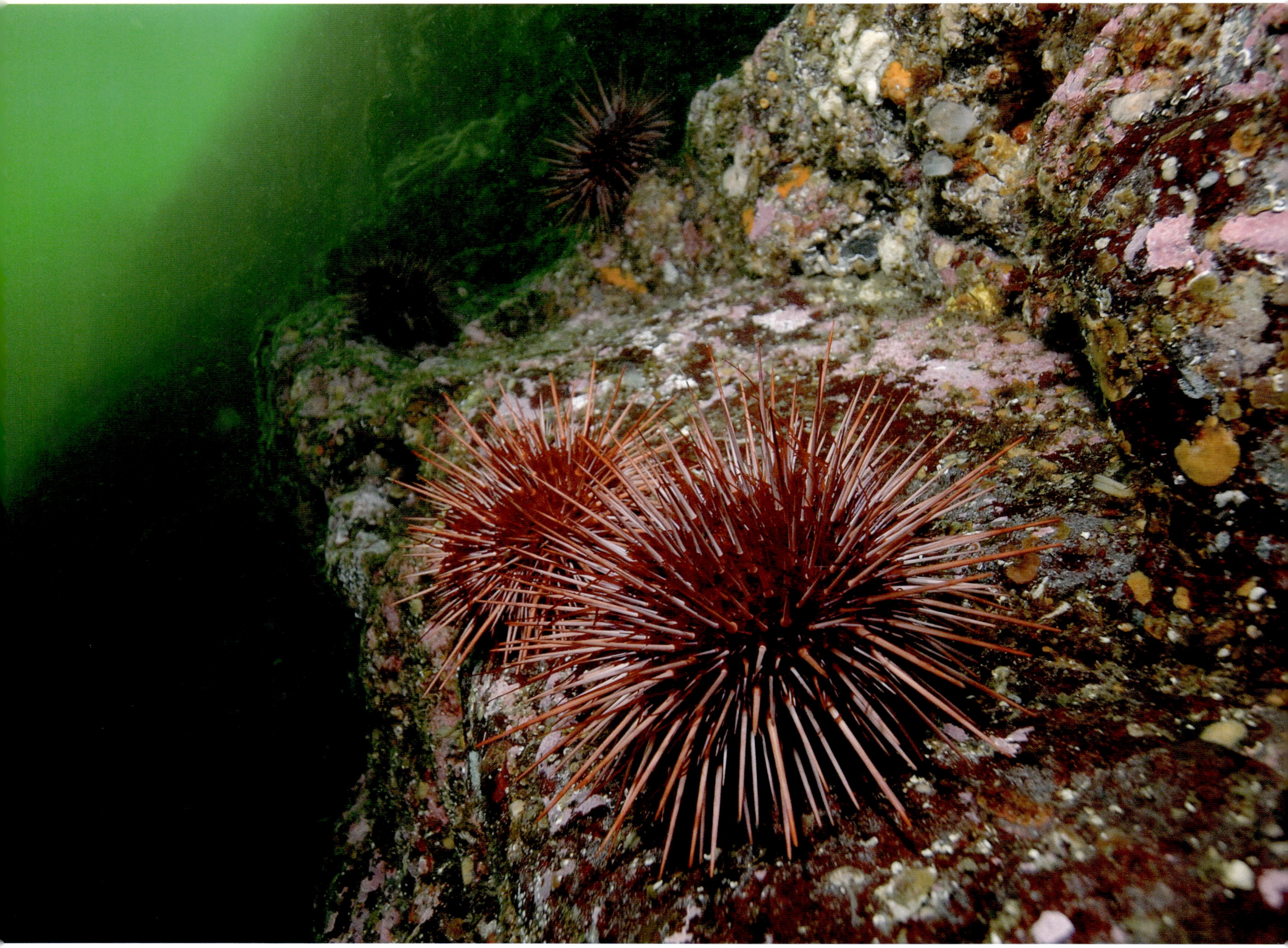

Sea Urchins, Browning Pass, Vancouver

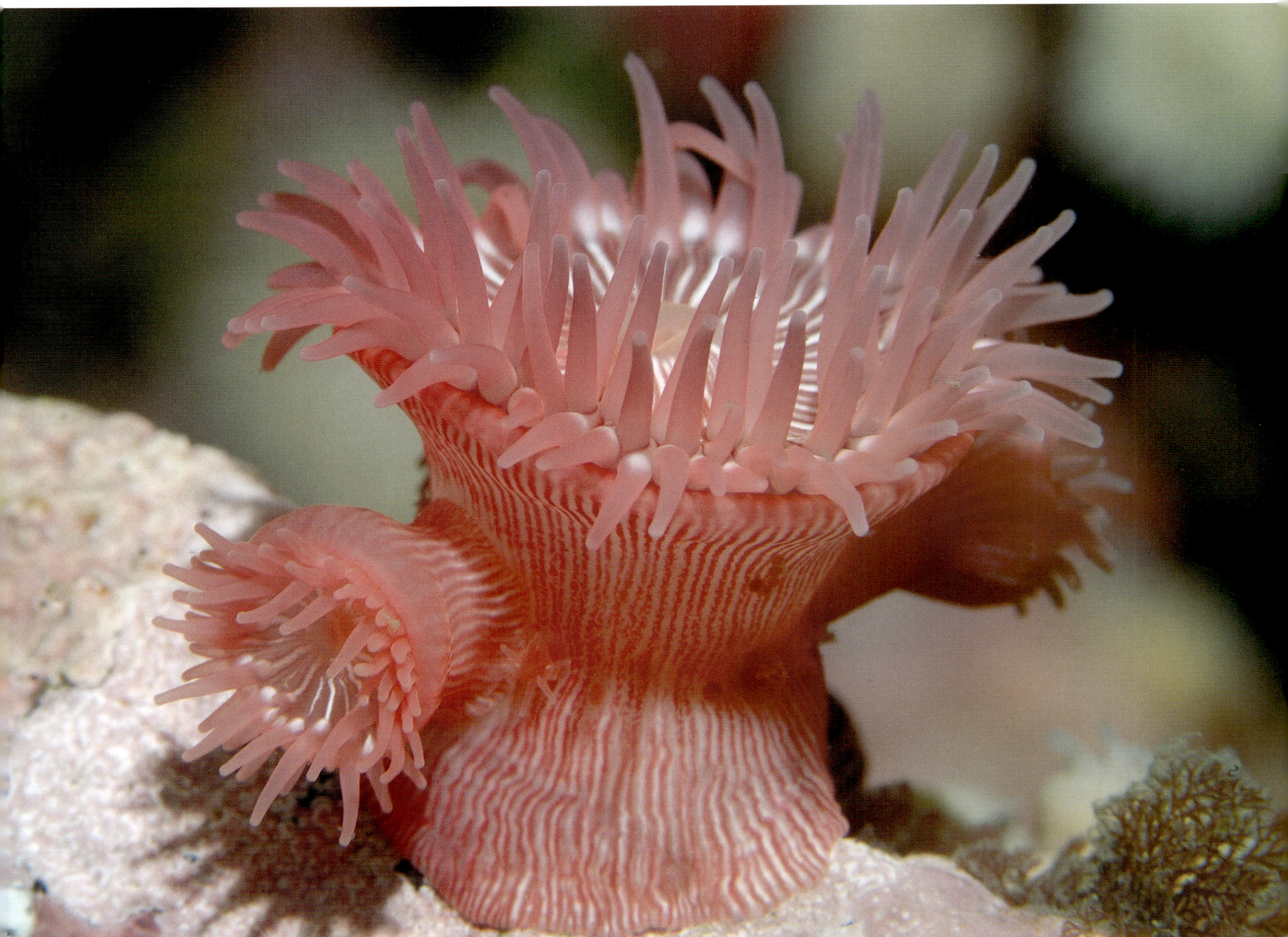

Brooding Anemones, Race Rocks, Vancouver

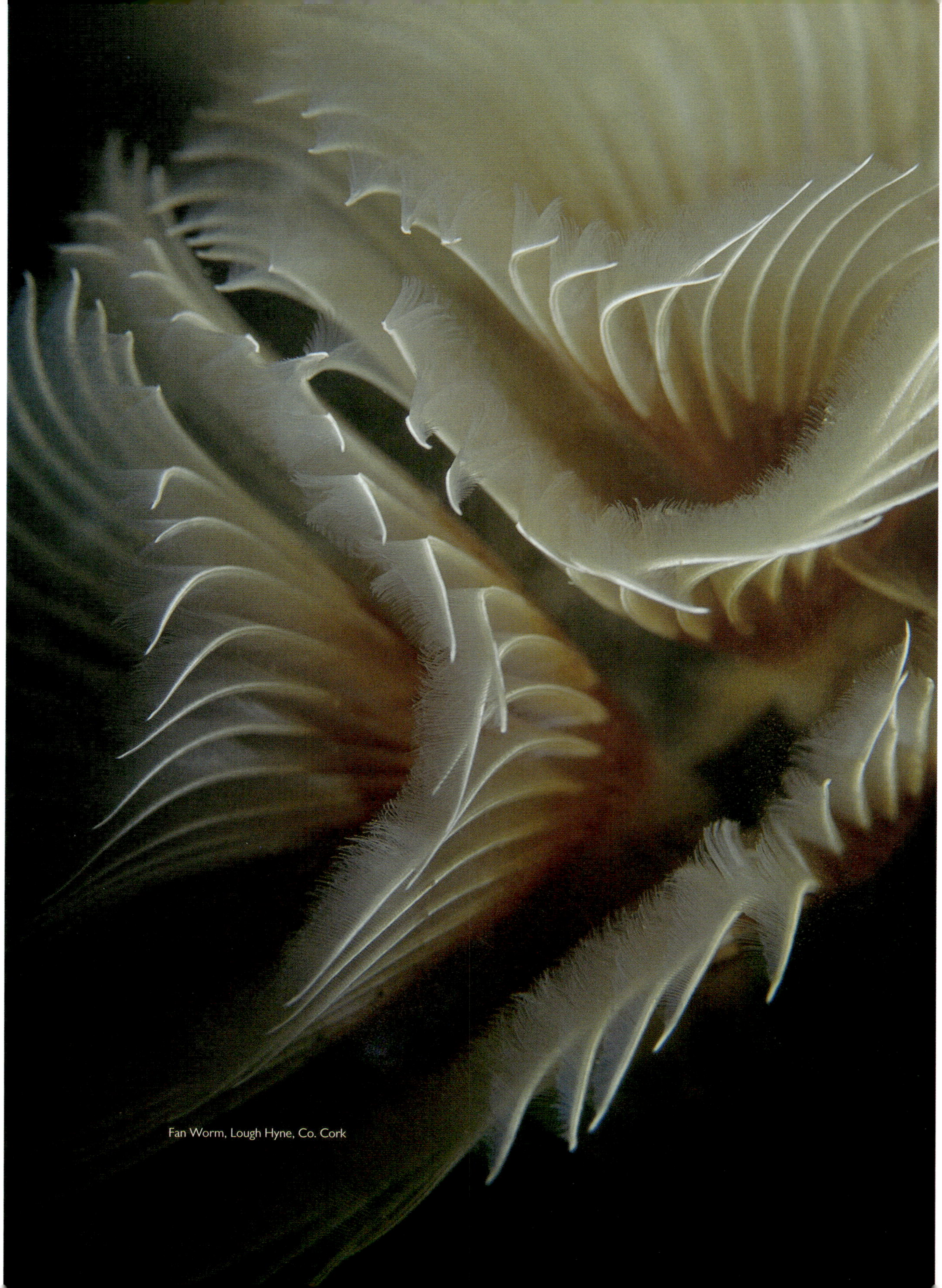

Fan Worm, Lough Hyne, Co. Cork

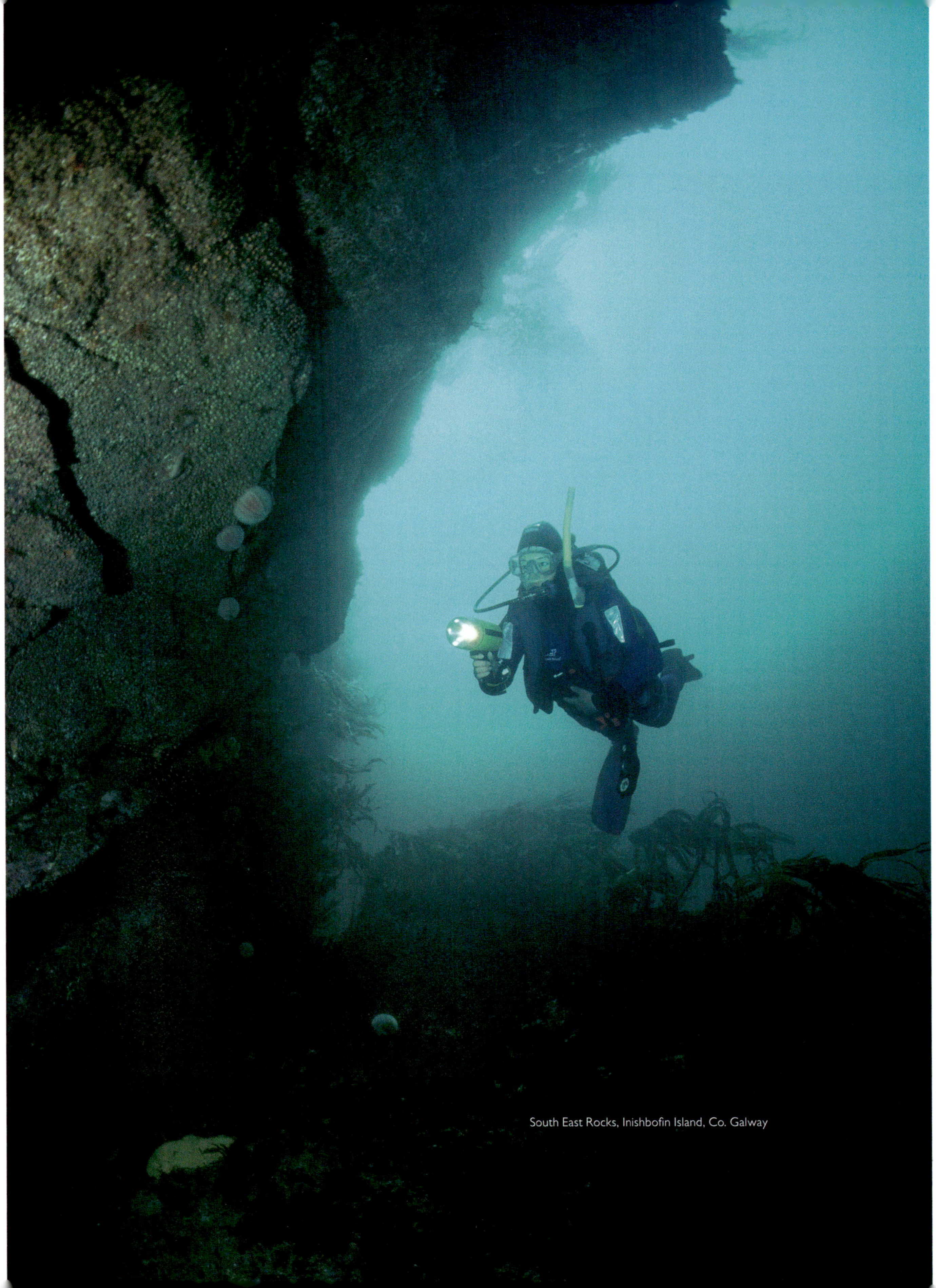

South East Rocks, Inishbofin Island, Co. Galway

Man stands with bowed head in the presence of nature's visible grandeurs, such as towering mountains, precipices, or icebergs, forests of immense trees, grand rivers, or waterfalls. He realizes the force of waves that can sweep away lighthouses or toss an ocean steamer about like a cork. In a vessel floating on the Gulf Stream one sees nothing of the current and knows nothing but what experience tells him; but to be anchored in its depths far out of the sight of land, and to see the mighty torrent rushing past at a speed of miles per hour, day after day and day after day, one begins to think that all the wonders of the earth combined can not equal this one river in the ocean.

J. E. Pillsbury, *The Gulf Stream*, 1891

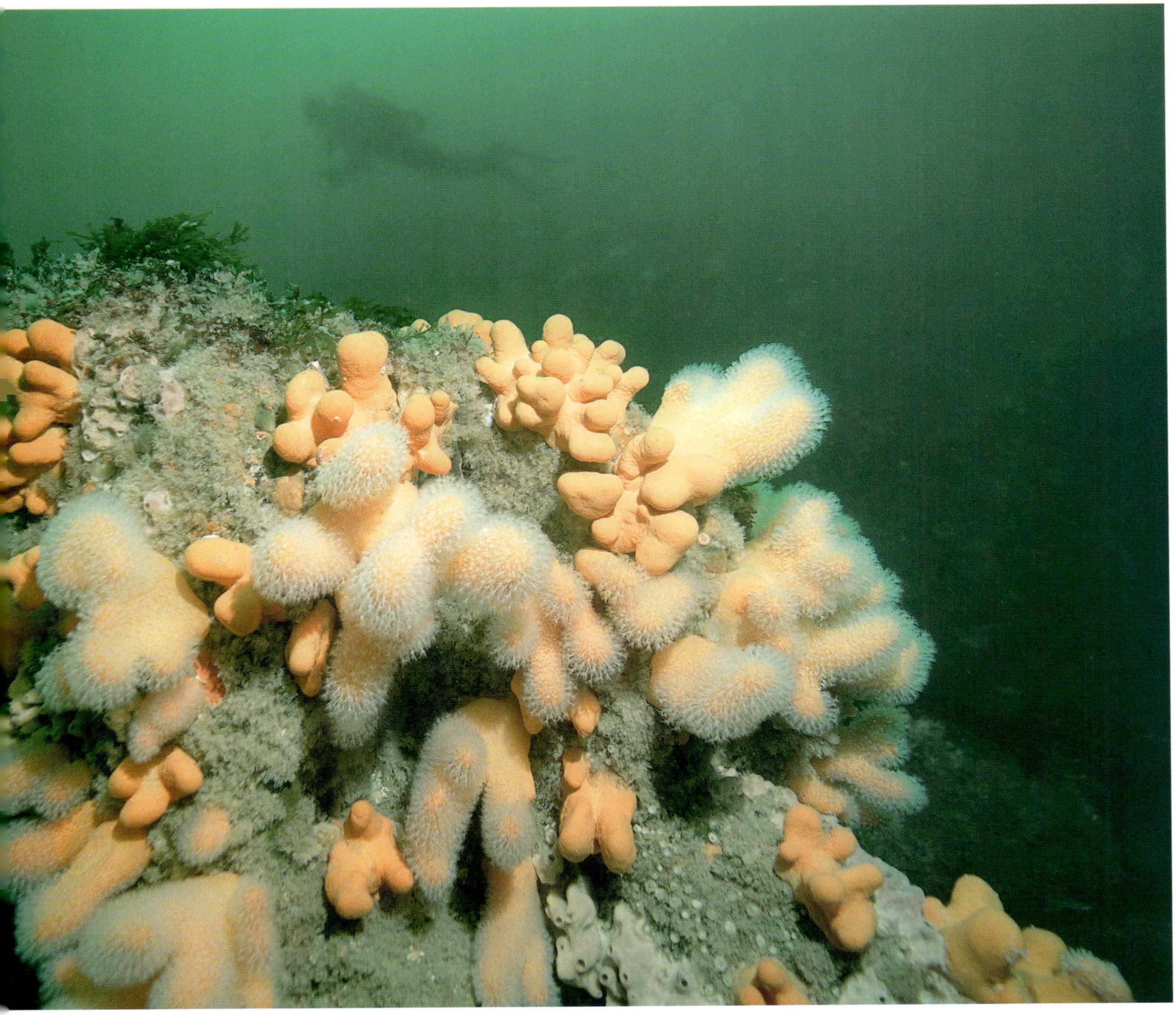

Rocky Reef, Inishmore, Aran Islands

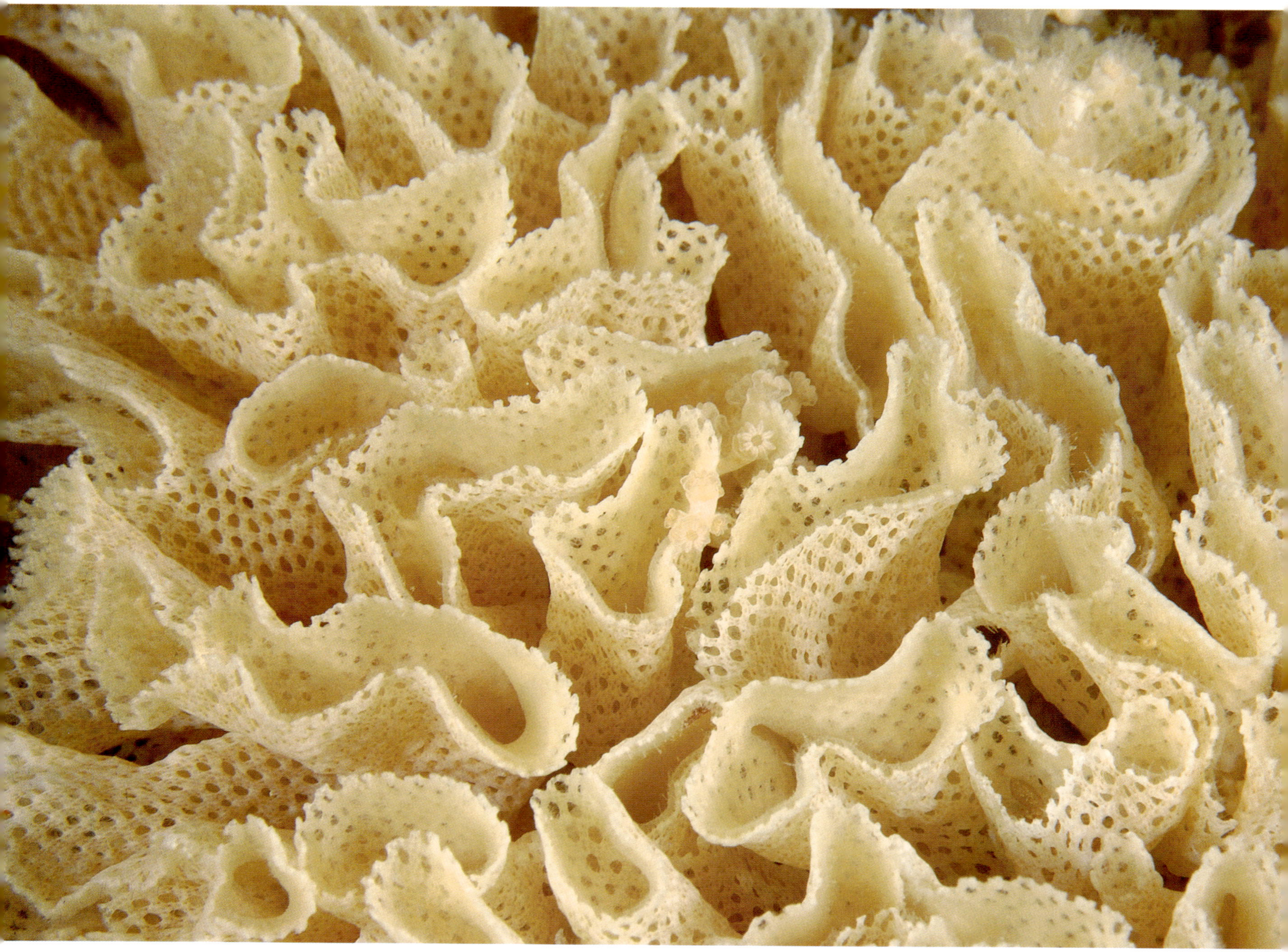

Lace Coral, Waterfall Bay, Tasmania

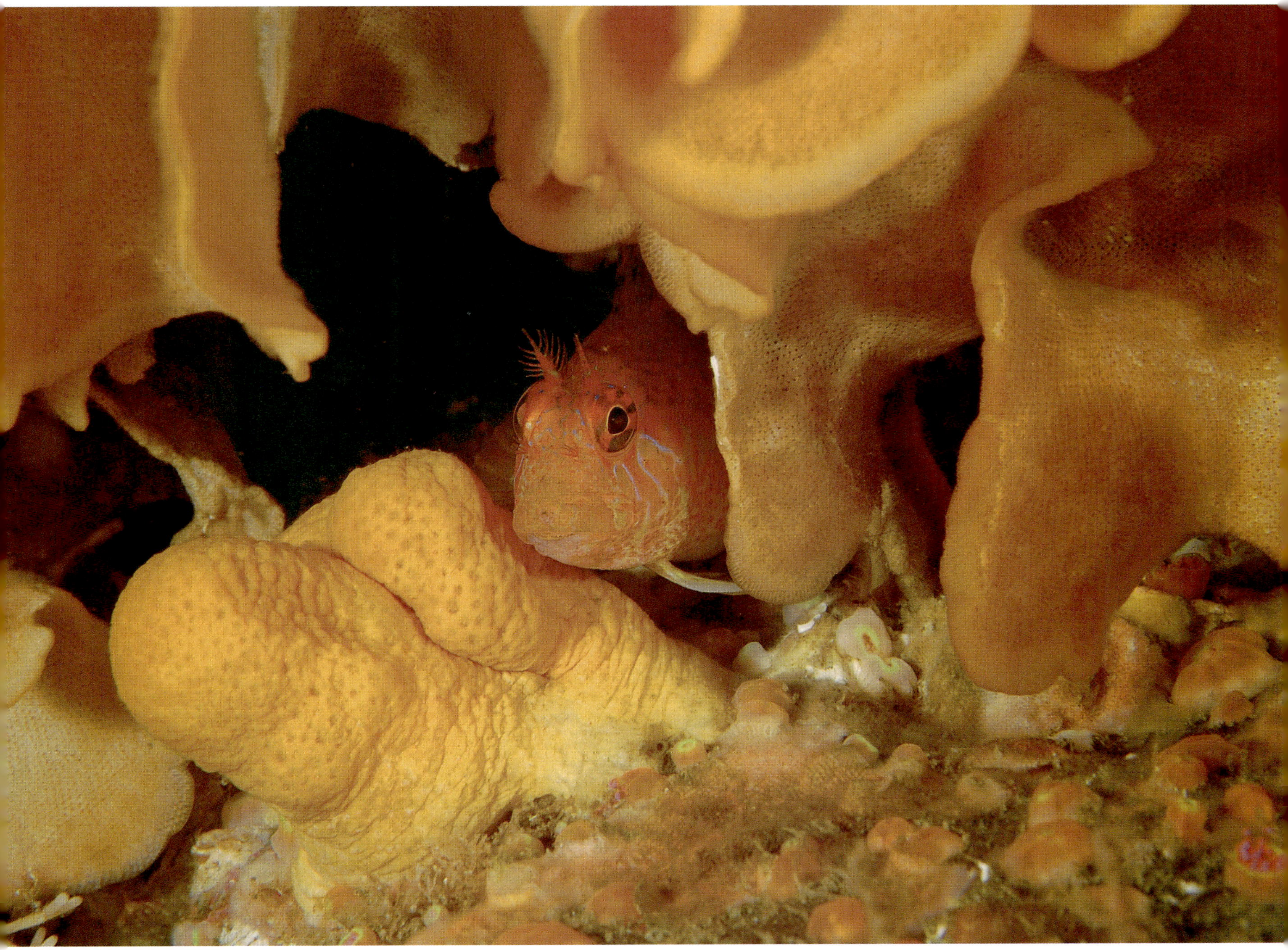

Red Blenny, St. Finian's Bay, Co. Kerry

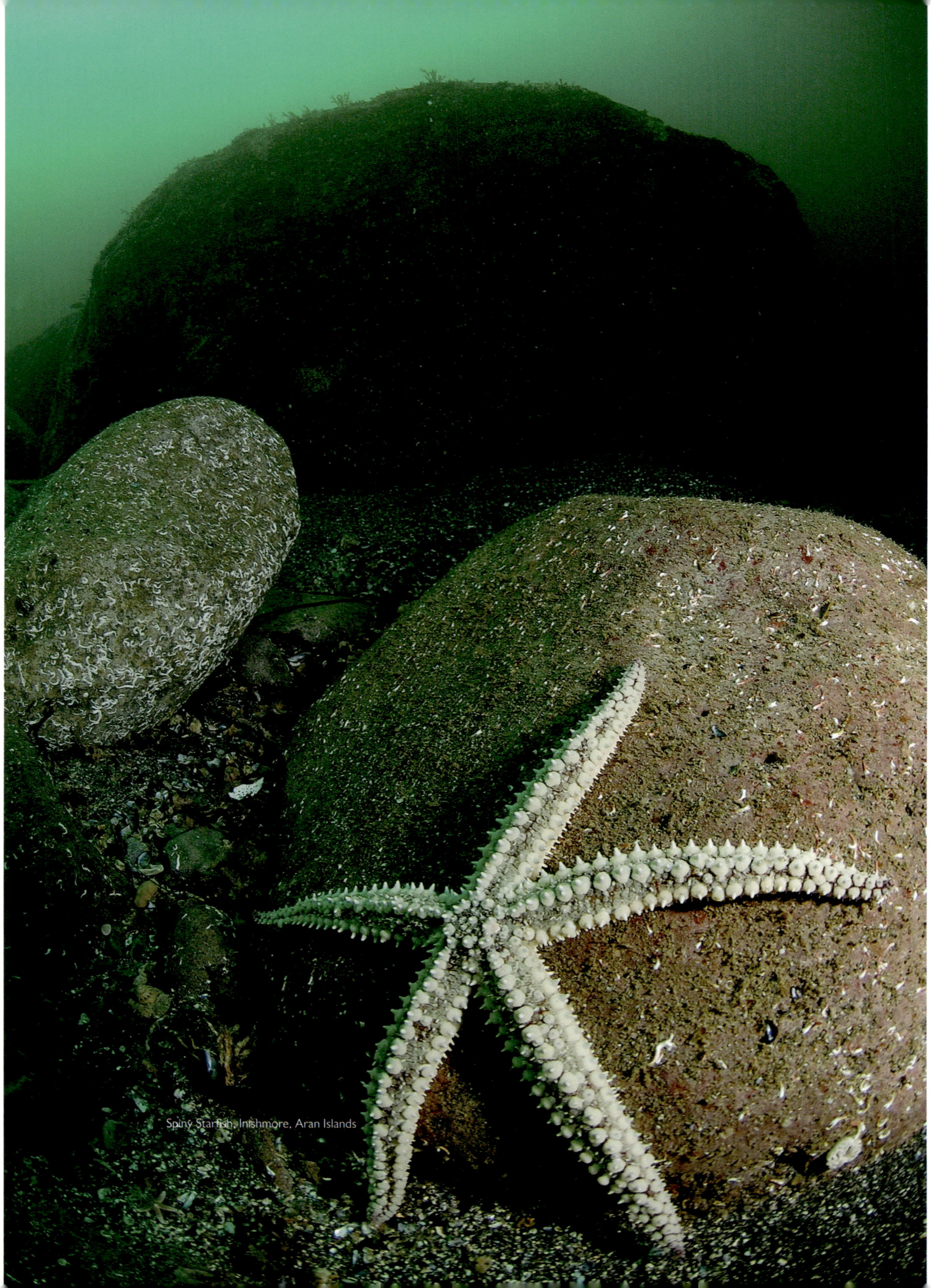

Spiny Starfish, Inishmore, Aran Islands

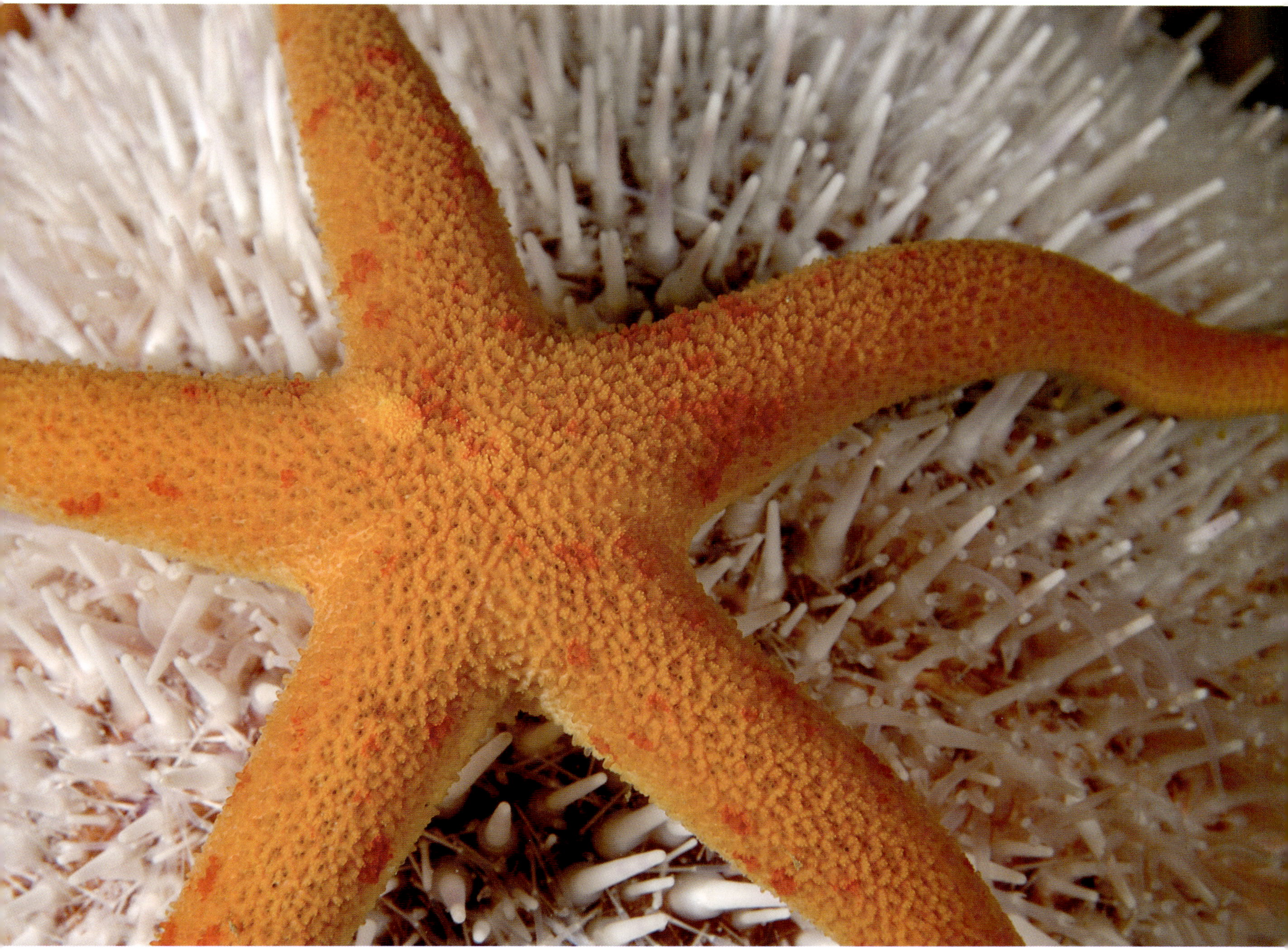

Starfish, St. Finian's Bay, Co. Kerry

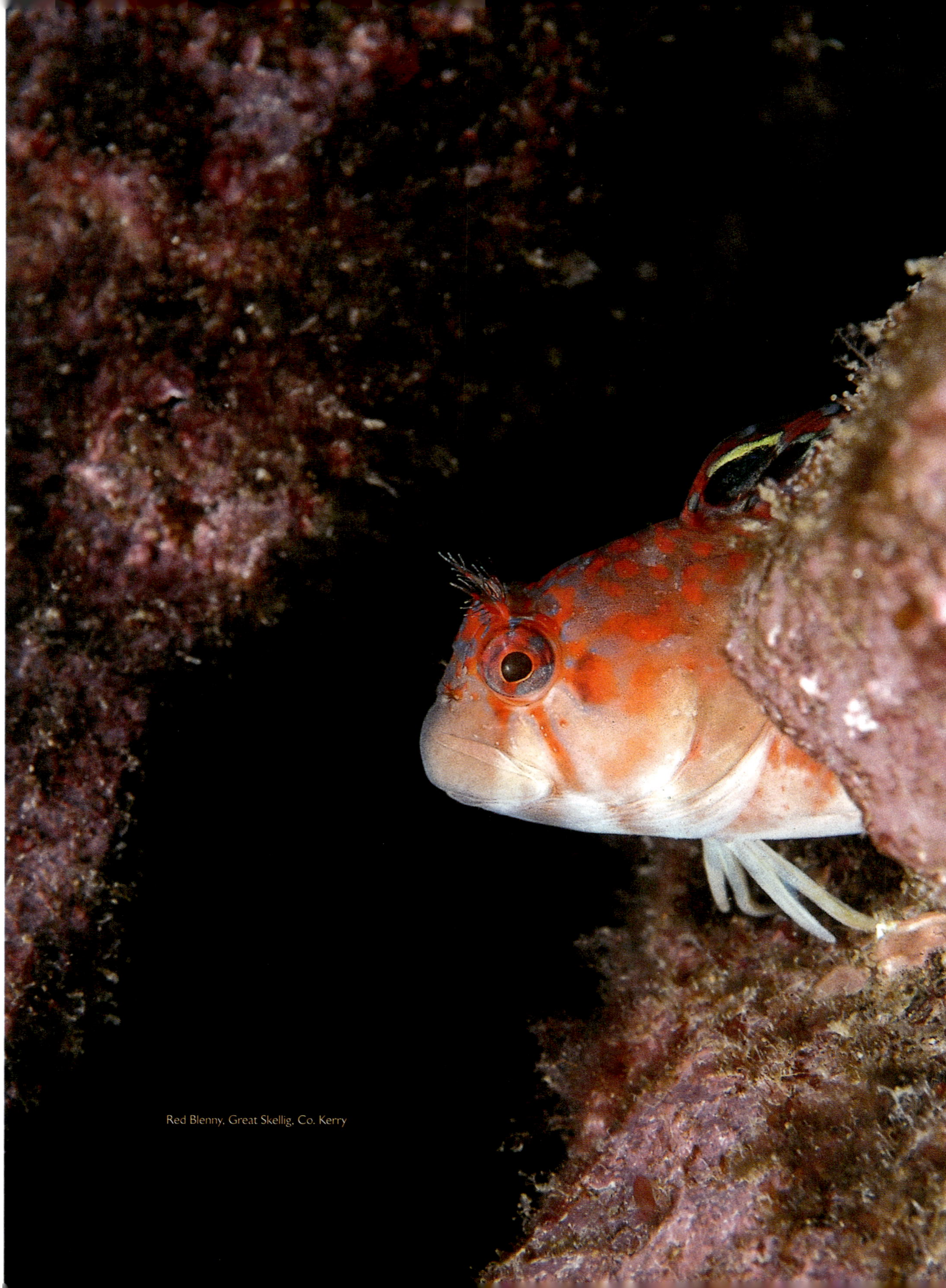
Red Blenny, Great Skellig, Co. Kerry

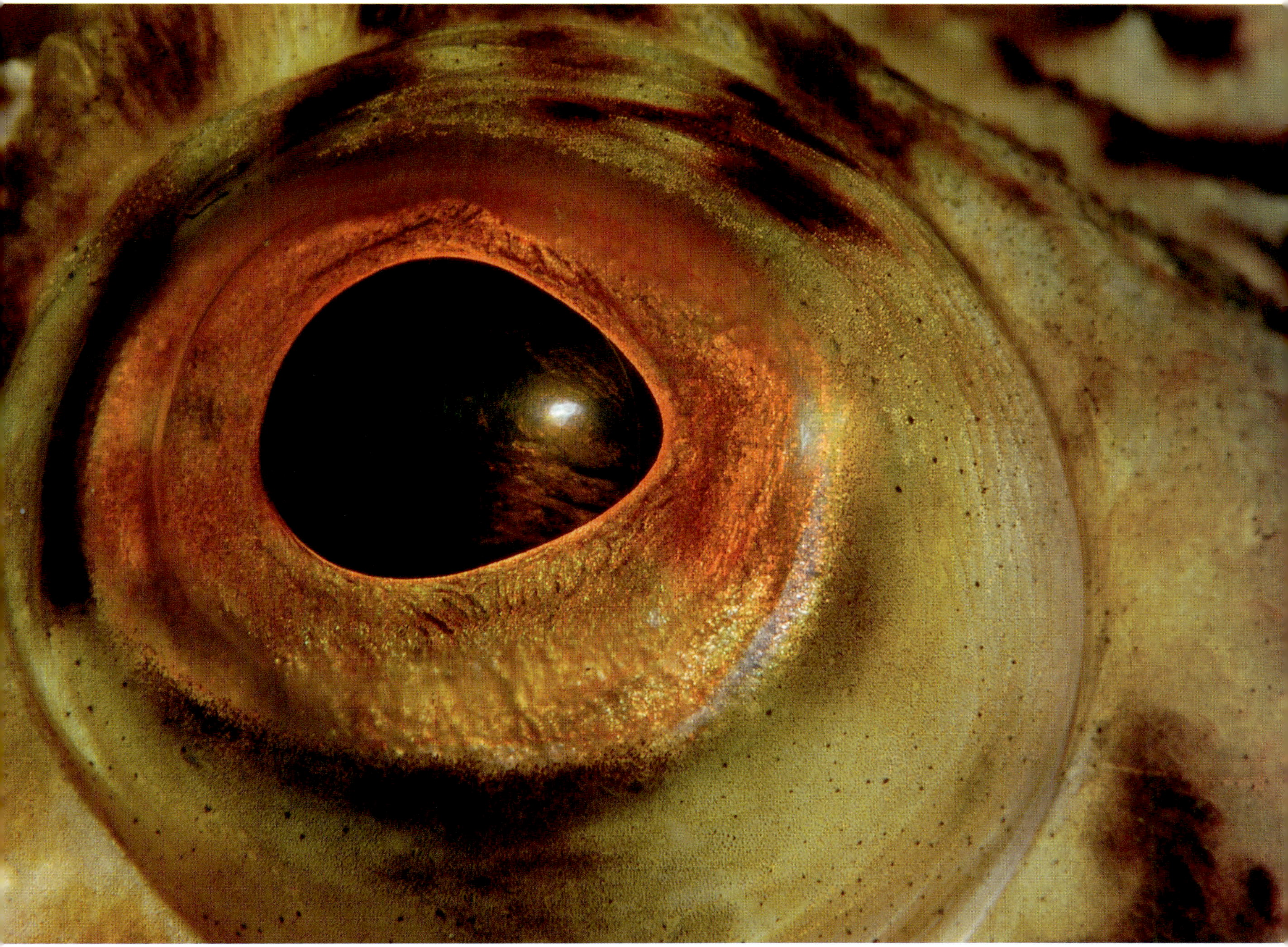

Sculpin Eye Detail, Race Rocks, Vancouver

In fact one does not go straight into the sea. Between the air and water, a steel blade quivers. What people call the surface is also a ceiling: a mirror from above, watered silk from beneath. Nothing is torn on the way through. Only a few bubbles mark the diver's channel and behind him the frontier soon closes. But once the threshold is crossed, one can turn back slowly and look up: that dazzling screen is the border between two worlds, as clear to one as to the other. Behind the looking glass the sky is made of water.

... Once he has broken the surface, the diver who is properly ballasted has no more weight, no more resistance: an aerial softness transports him where he wills. Here the world is sweetness. There is not a place in his body, from head to foot, which is not relaxed. It is a pleasure to stretch out, to lie on one's back and to feel the perfect fluency of one's muscles. Dreams float very slowly up from the sea. Walled in silence and completely alone, the diver begins an interior monologue in the cell of his undreamed-of content.

Philippe Diolé, *The Undersea Adventure*, 1953

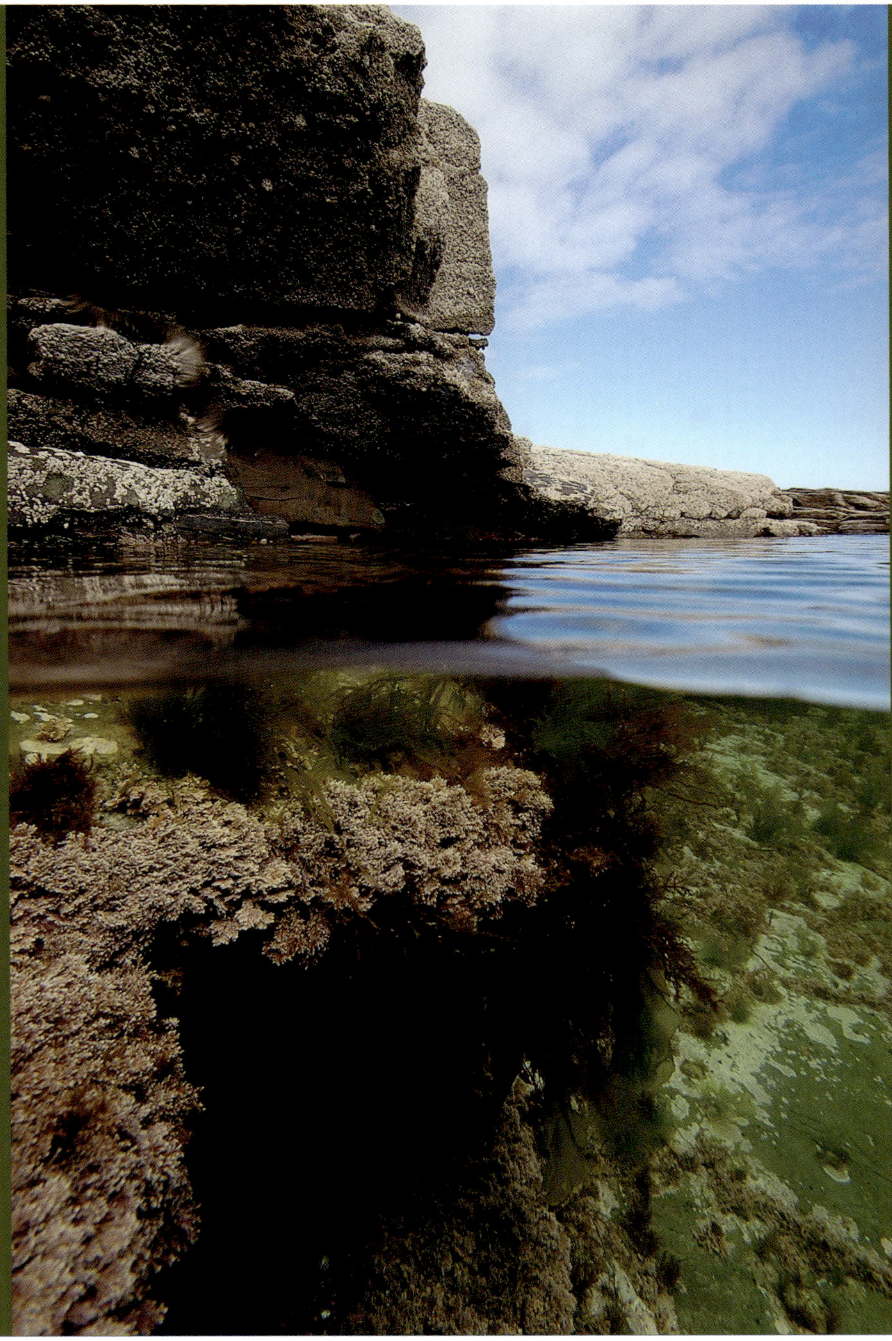

Pollack Holes, Kilkee, Co. Clare

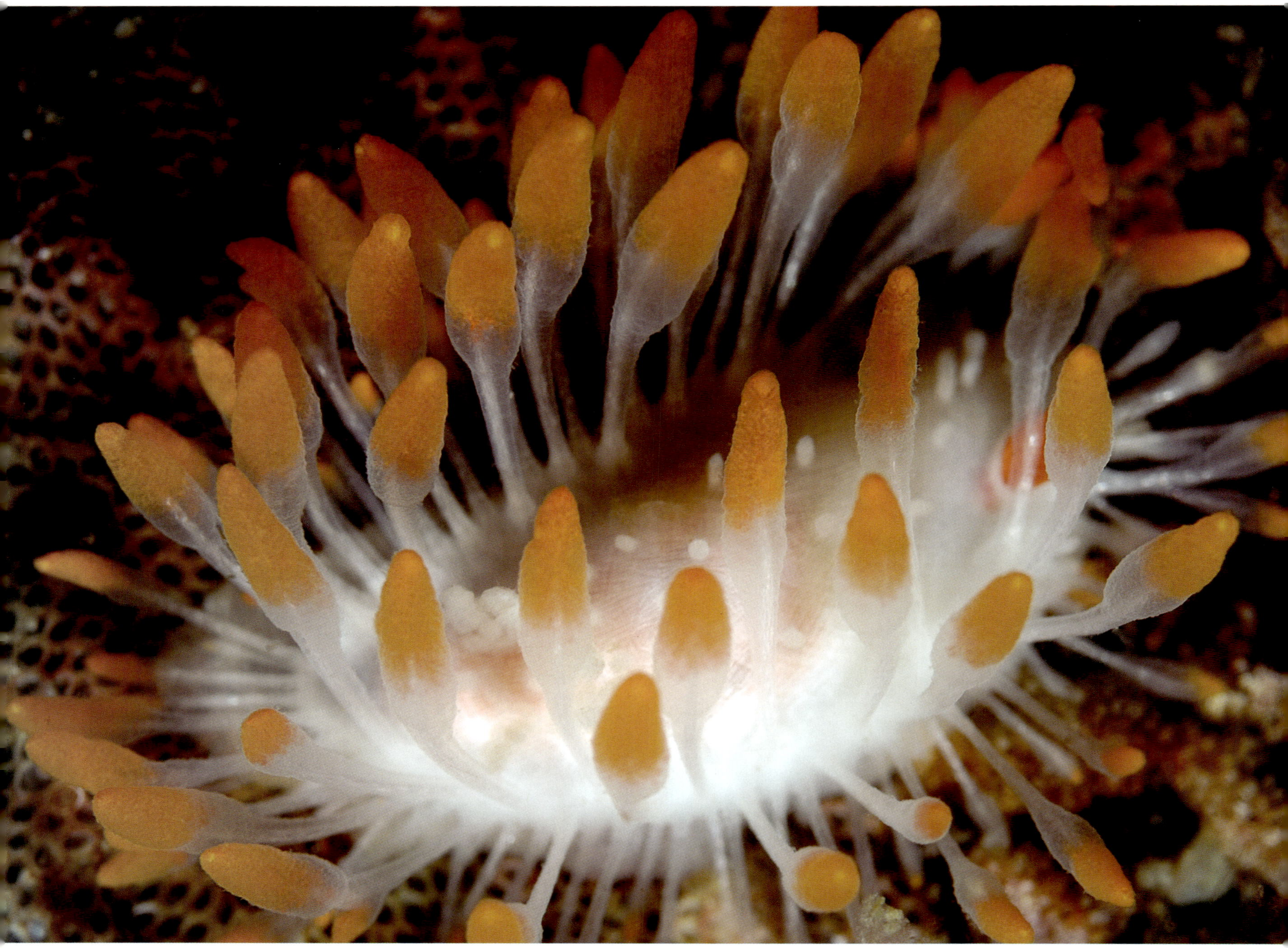

Orange Tipped Nudibranch, Browning Pass, Vancouver

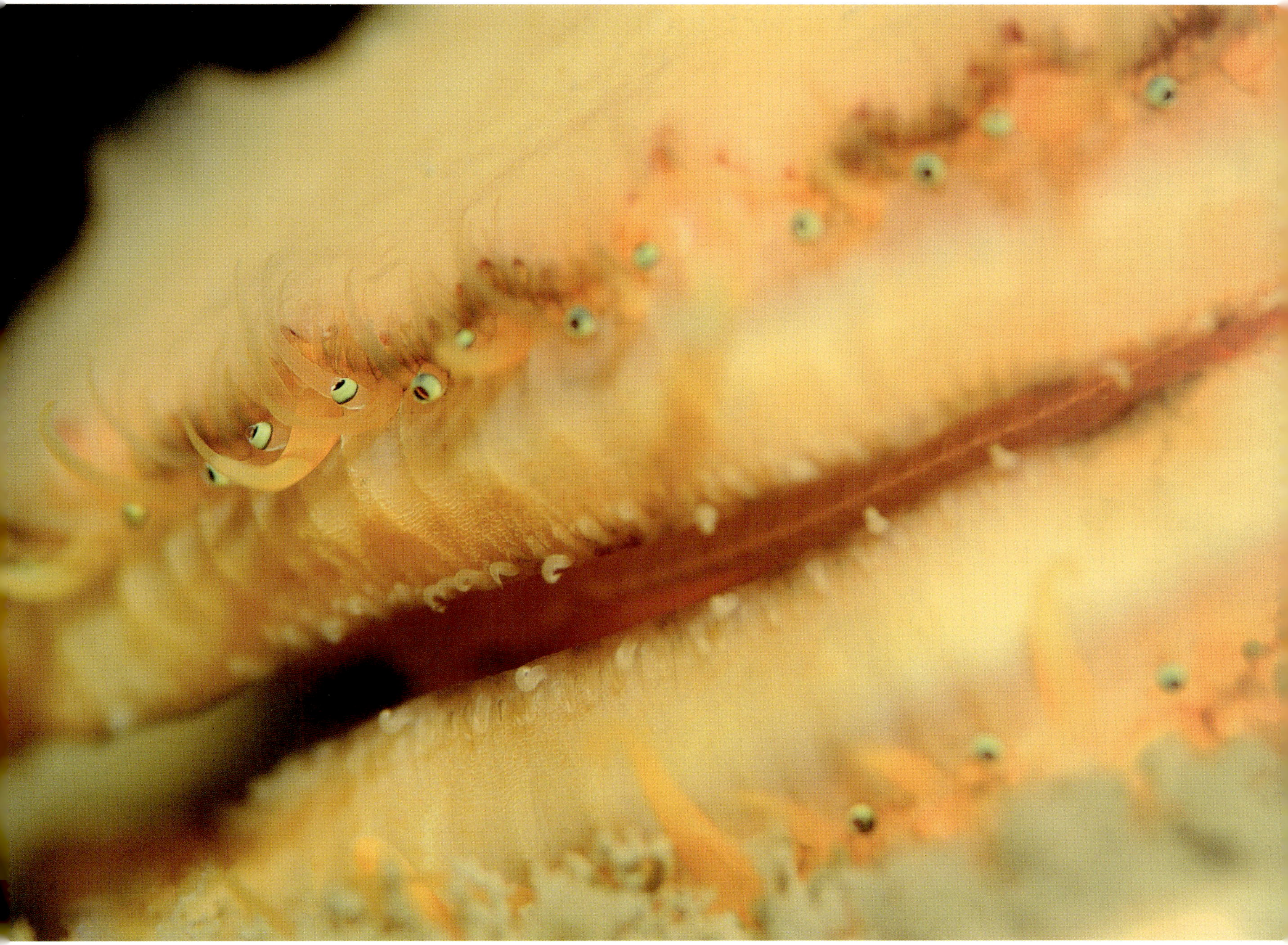

Scallop Eyes, Ogden Point, Vancouver

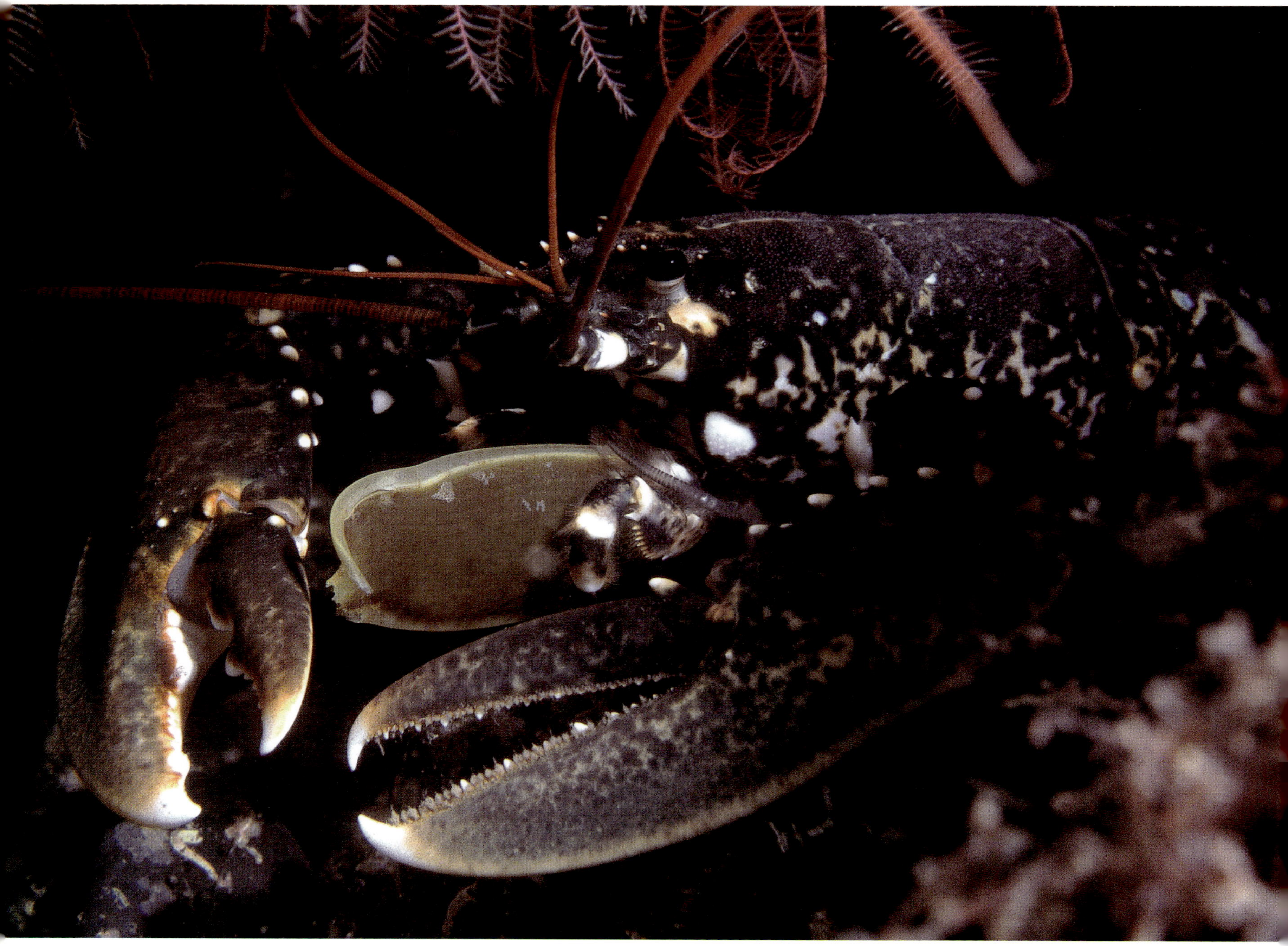

Lobster, Eating Mermaid's Purse, Old Head of Kinsale, Co. Cork

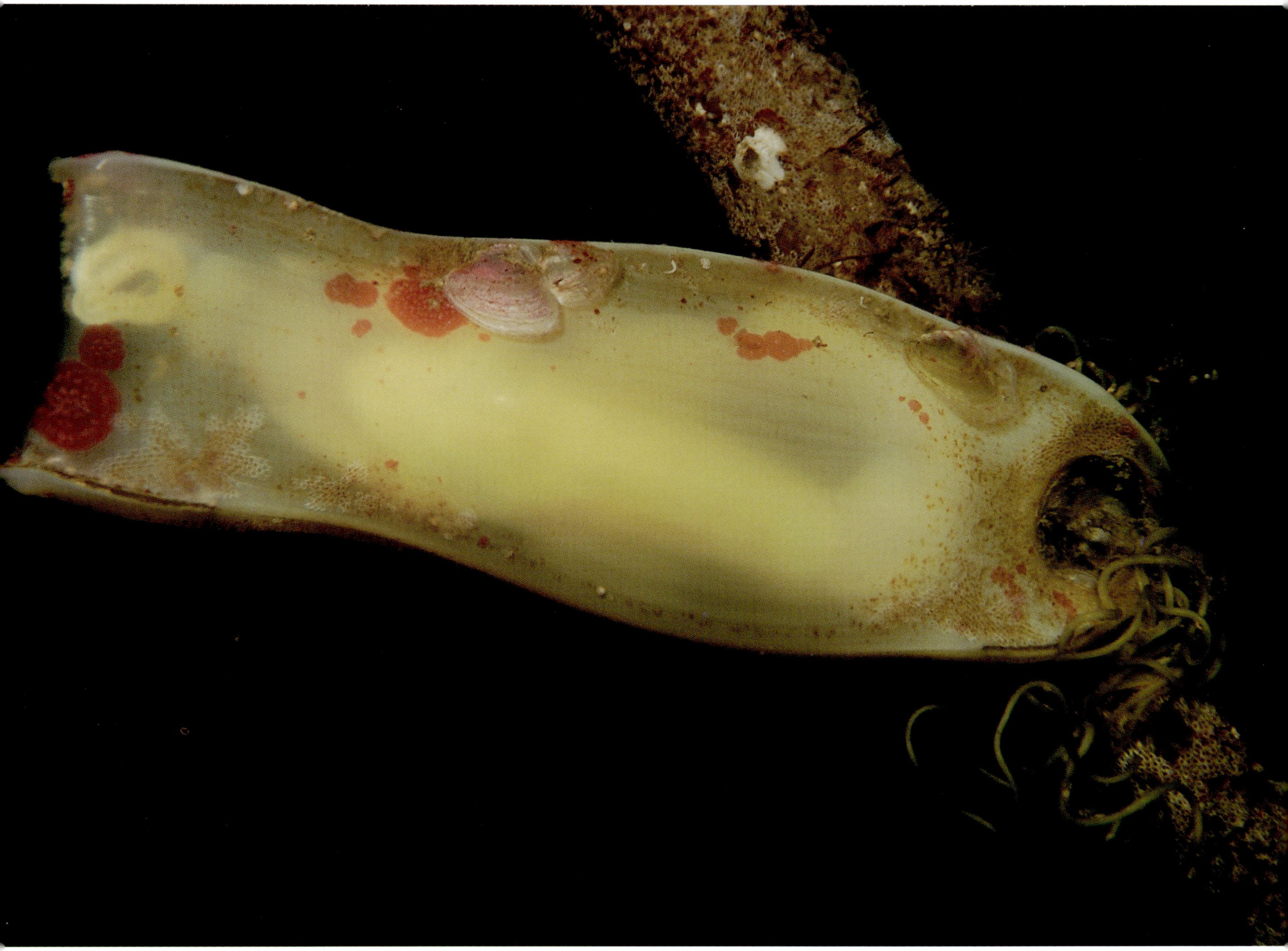

Juvenile Dogfish in Mermaid's Purse, Kenmare Bay, Co. Kerry

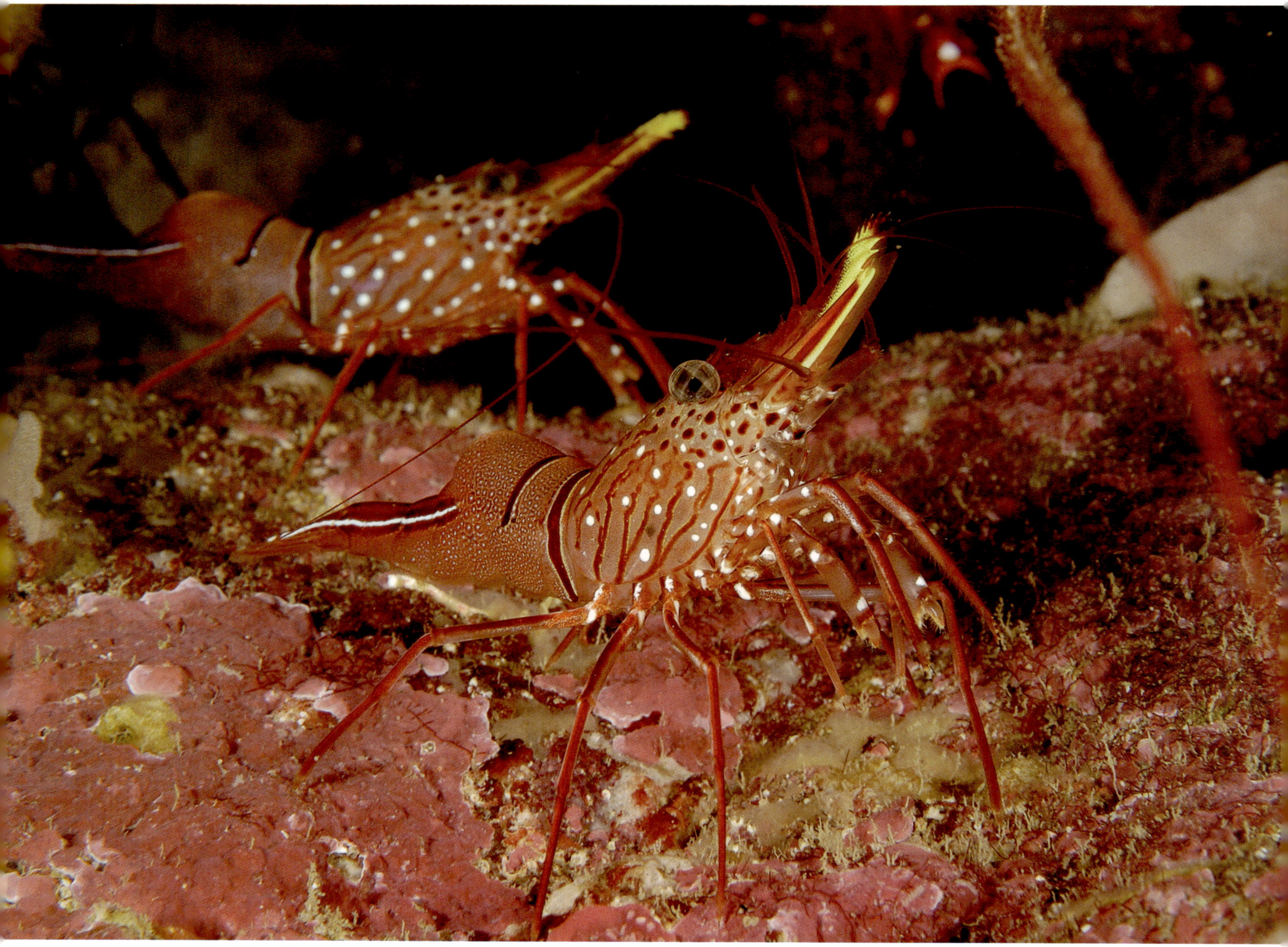

Shrimps, Waterfall Bay, Tasmania

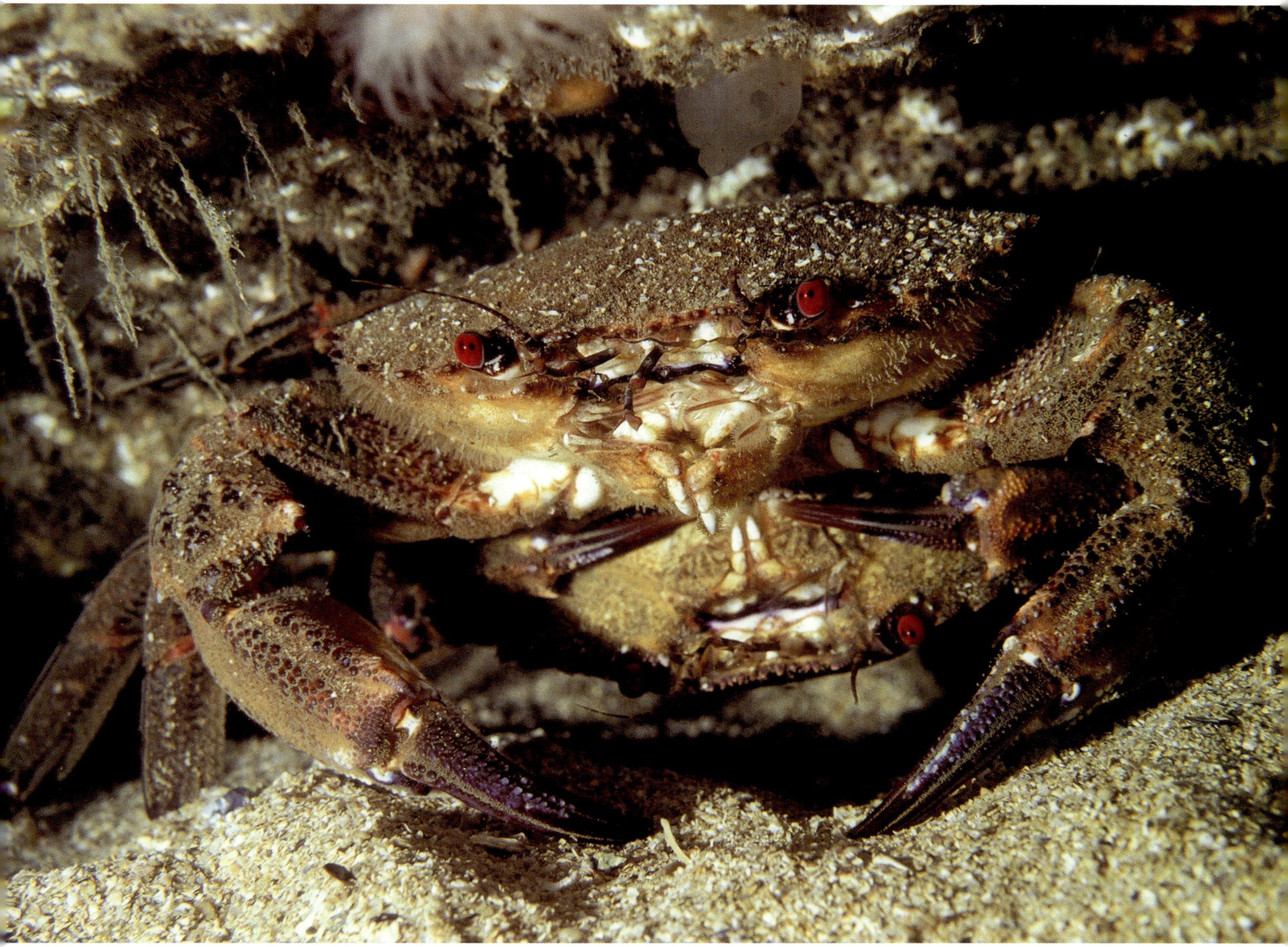

Velvet Swimming Crabs, Old Head of Kinsale, Co. Cork

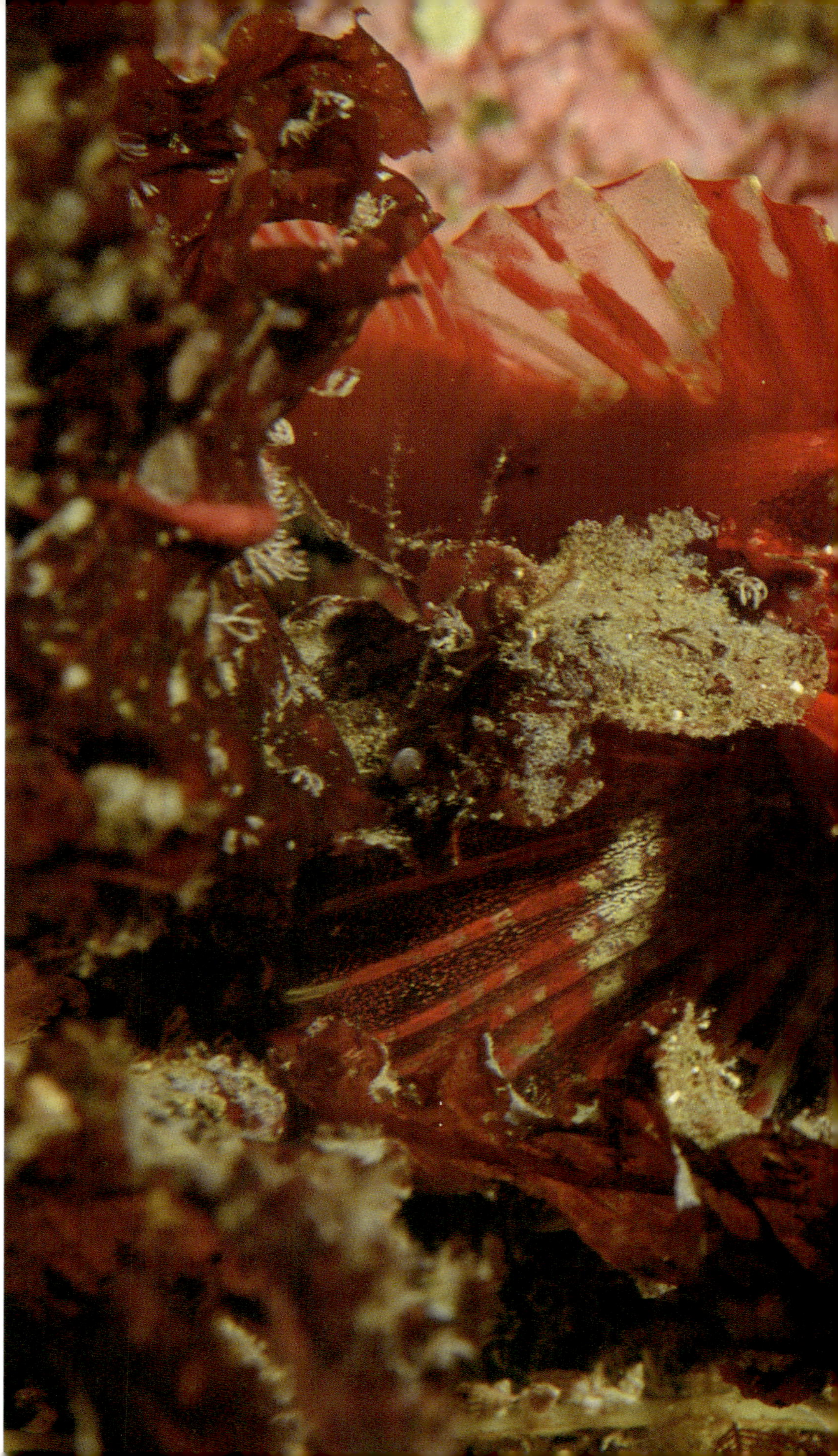

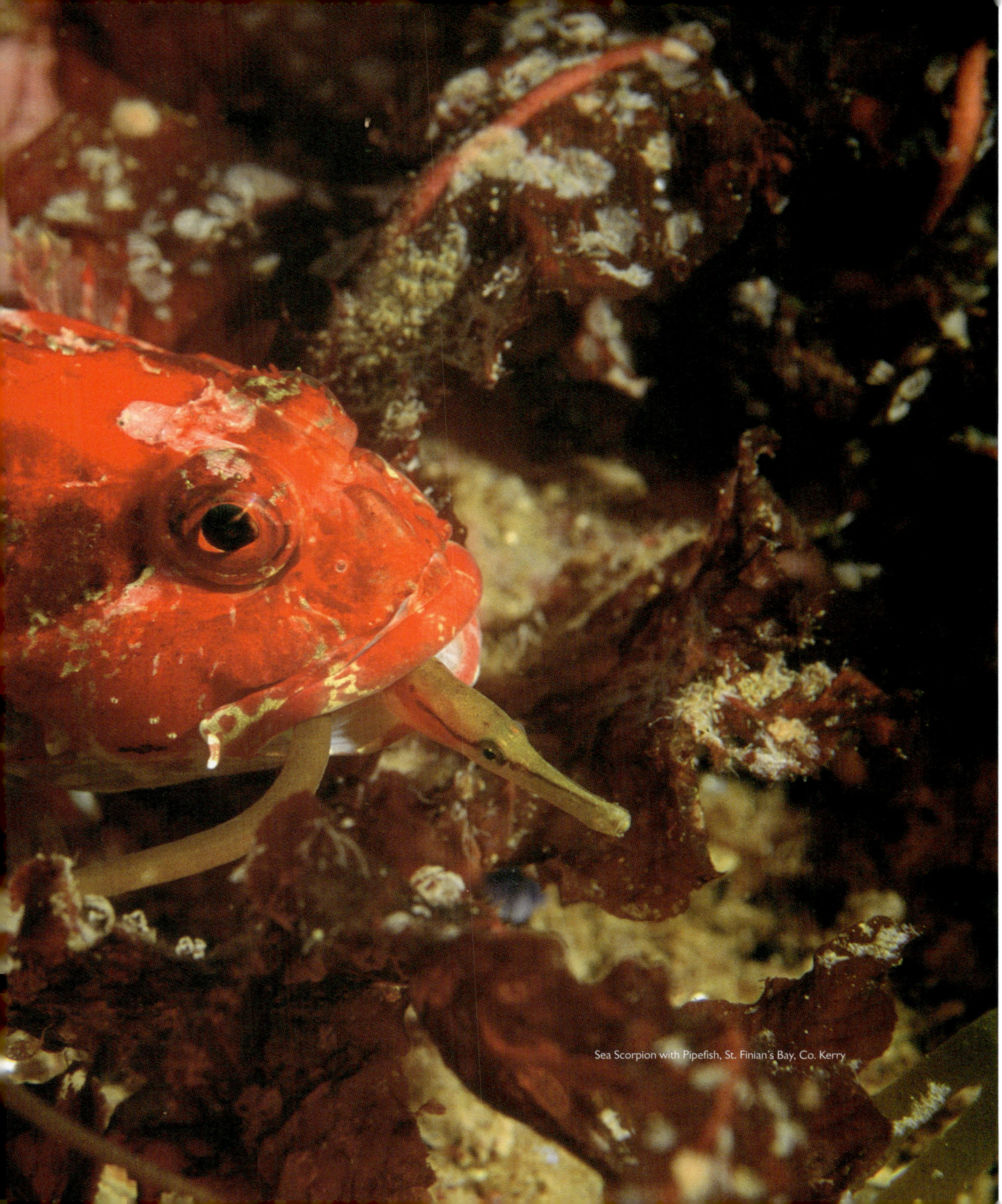
Sea Scorpion with Pipefish, St. Finian's Bay, Co. Kerry

Eventually man, too, found his way back to the sea.
... over the centuries, with all the skill and ingenuity and reasoning powers of his mind, he has sought to explore and investigate even its most remote parts, so that he might re-enter it mentally and imaginatively.

... And yet he has returned to his mother sea only on her own terms. He cannot control or change the ocean as, in his brief tenancy of earth, he has subdued and plundered the continents. In the artificial world of his cities and towns, he often forgets the true nature of his planet and the long vistas of its history, in which the existence of the race of men has occupied a mere moment of time. The sense of all these things comes to him most clearly in the course of a long ocean voyage, when he watches day after day the receding rim of the horizon, ridged and furrowed by waves; when at night he becomes aware of the earth's rotation as the stars pass overhead; or when, alone in this world of water and sky, he feels the loneliness of his earth in space. And then, as never on land, he knows the truth that his world is a water world, a planet dominated by its covering mantle of ocean, in which the continents are but transient intrusions of land above the surface of the all-encircling sea.

Rachel L. Carson, *The Sea Around Us*, 1951

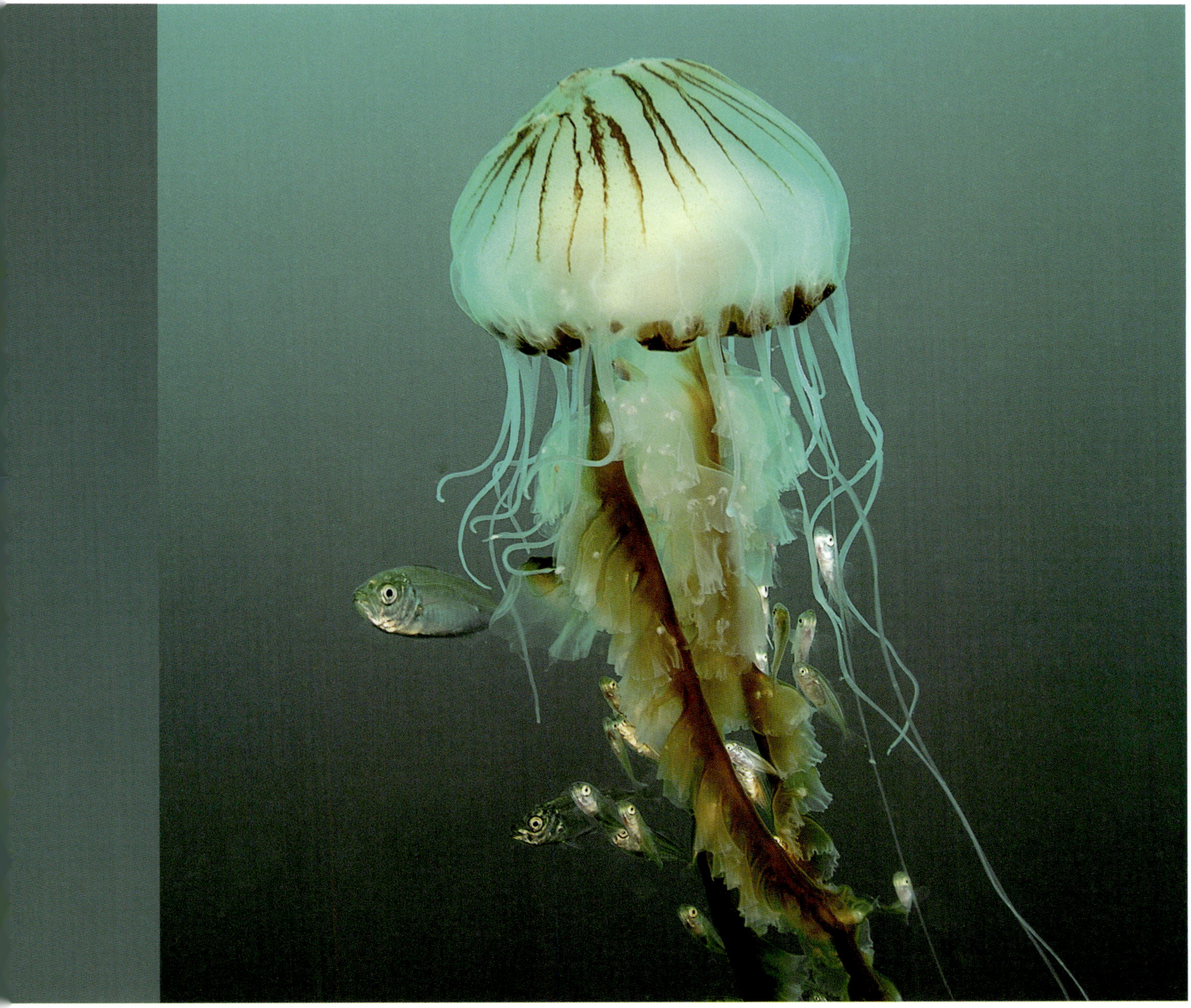

Compass Jellyfish and Juvenile Whiting, St. Finian's Bay, Co. Kerry

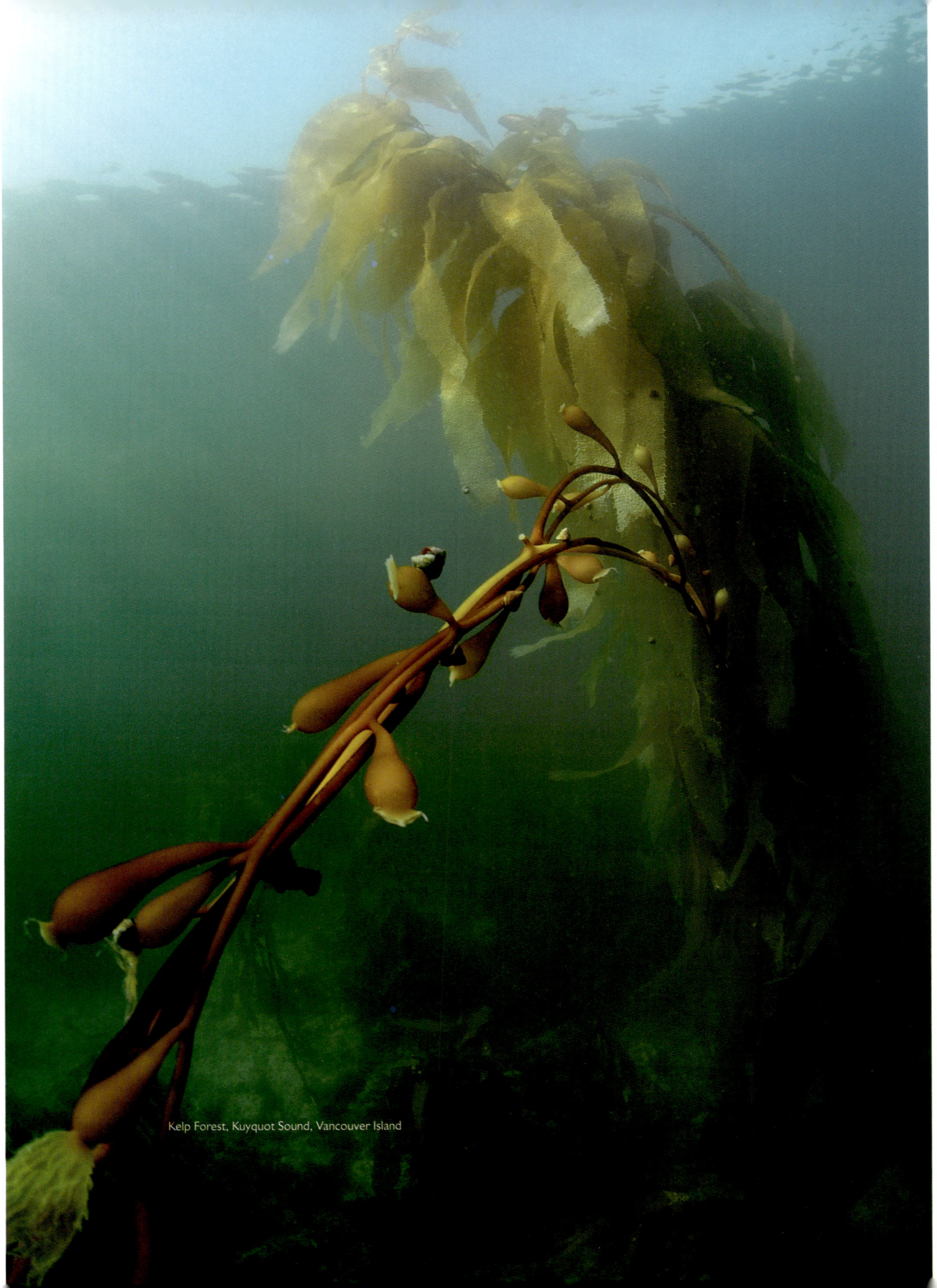

Kelp Forest, Kuyquot Sound, Vancouver Island

THE OCEAN'S RHYTHMS

I can only compare these great aquatic forests of the southern hemisphere with the terrestrial ones in the intertropical regions. Yet if in any country a forest was destroyed, I do not believe nearly so many species of animals would perish as would here, from the destruction of the kelp. Amidst the leaves of this plant numerous species of fish live, which nowhere else could find food or shelter; with their destruction the many cormorants and other fishing birds, the otters, seals, and porpoises, would soon perish also.

Charles Darwin,
The Voyage of the Beagle, 1909

Winter diving can be surprisingly rewarding, despite the cold. On a frosty first dive of the year, the unusually clear conditions are a little strange. Swimming slowly to depth off the Old Head of Kinsale, the water seems quiet, like a holiday park in mid-winter. Sea grass and kelp waft gently to the ocean's rhythm. Everybody seems to have left, except for a few small crabs scurrying comically sideways out of my path. A scallop shell senses approaching danger and suddenly jumps from the sandy bottom, pumping its way to escape. I look around for the usual fish life, but all have gone deep or migrated for winter.

The ocean's cool seas change dramatically with the seasons. While we rarely think of winter as being beneficial, it is a time of rejuvenation for the sea. The water begins to cool and the top layer, which has been warmed throughout the summer months, begins to mix with cooler water at depth. Torrential rain and strong winds contribute to the health of the seas by adding nutrients and mixing these water layers. By the time the clocks are adjusted for daylight saving, the phytoplankton are already blooming and the explosion of life begins once more. Right across the oceans, this seasonal rhythm of boom and bust characterises the temperate seas.

Water temperatures vary with oceanic currents, some seas getting much colder than others. In western Europe, the Gulf Stream prevents the water from cooling much below ten degrees Celsius, whereas a chilly six degrees is common along the Pacific coast of North America. The cold here is invigorating, not just to the intrepid diver but to the life in these rich seas. The Gulf Stream brings oxygen and nutrients from the deep ocean that are distributed by up-welling currents and tidal streams along the coasts, from Alaska to California. Vancouver Island, off Canada's British Columbia, is bathed by particularly lush emerald seas. Beneath its forest-covered coastline the cold water bites the exposed skin of the face and the first few metres of water are dense green. Then, at depth, the water clears and a darkened landscape emerges. Walls of bright yellow cloud sponges appear in the dive lights and a whole community life is revealed. Much of it is similar to that seen in other cool seas, but the size of everything is much larger; the super-rich plankton giving everything a body-building boost. It is most noticeable in areas of strong tides, as billions of litres of water are channelled through the narrows and islets twice daily. The invertebrate animal life blankets every surface and whole walls are completely colour-washed with life.

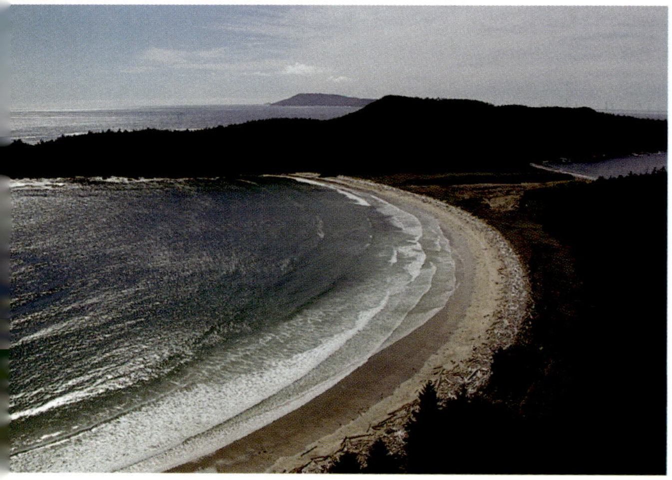

Cape Scott, Vancouver

Advances in scuba gear have brought about great changes in how we can visit even the coldest waters. The rebreather, a forerunner to the popular aqua-lung, has made a comeback in modern times. By re-using the divers breathing gas, dive times can be greatly extended. The exhaled breath is scrubbed of carbon dioxide with soda lime, a process which generates heat. The enriched gas is thus warmed and prevents heat loss through the lungs. This comfort, along with that from modern insulating fabrics worn under the dry-suit, allows dives of an hour or more even in the coolest waters. These advances allow us to explore our temperate seas year-round – winter time offering the best opportunity to see them at their clearest.

As winter descends on the Southern Ocean, the 'roaring forties' become the wildest seas on earth. But in the shelter of the Tasman Peninsula's grand cliffs, the waters are enchanting underwater gardens and home to some bizarre residents. On my first dive here, I am amazed at the clarity of the water while descending into a shallow area of kelp. I am hoping to see some weedy sea-dragons, slow moving relatives of sea-horses. Growing much larger, to about forty centimetres, they live only in the seas off southern Australia. Within minutes, I see a female, golden brown with zebra stripes splashed with yellow and purple. She is unlike anything else in the ocean and much larger than her male counterpart. In a role reversal rare in the animal kingdom, it is the male who carries the eggs and gives birth. Twice during each breeding season, the female lays 100–250 eggs onto a special 'brood patch' on the underside of the male's tail, where they are attached and fertilised. The bright pink eggs develop for about six weeks after which they hatch as miniature adults. The yolk sac sustains them for a couple of days but then they must fend for themselves among the kelp. This world of the kelp forest is one of diving's more bizarre experiences.

Dropping beneath the surface, surrounded by huge plants that reach all the way to the sea-bed, it is a free-fall unlike any other. Half way down, I add some air to my buoyancy jacket and halt the descent. With a few fin kicks, I begin to glide through the forest, sunlight flickering through the kelp fronds, shimmering on the sandy bottom below. It is like being suspended in a tropical rainforest. Here, in Fortesque Bay, I am in one of the few remaining kelp forests in Tasmania. They have declined steadily over the past thirty years, for reasons we have yet to fully

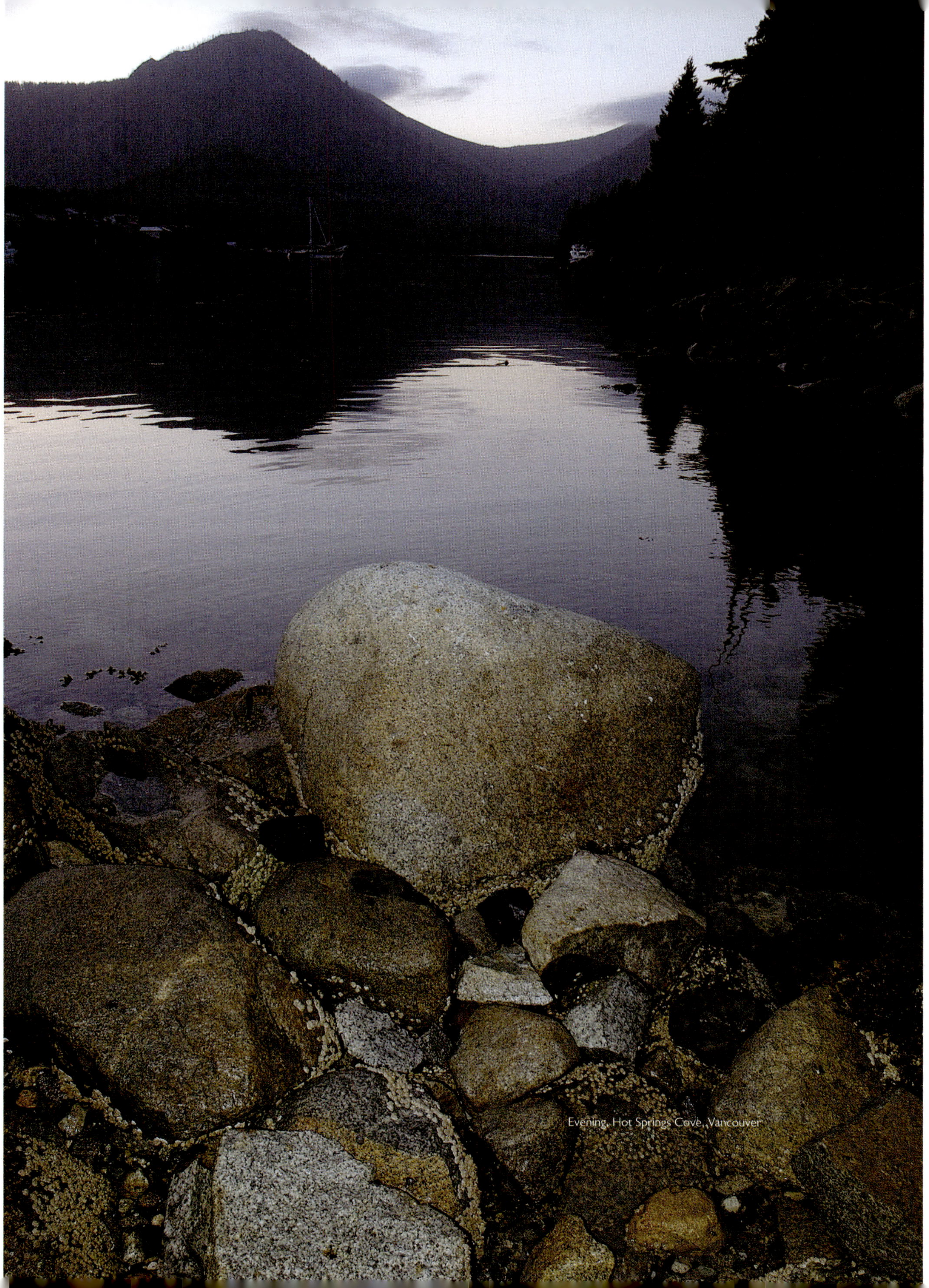

Evening, Hot Springs Cove, Vancouver

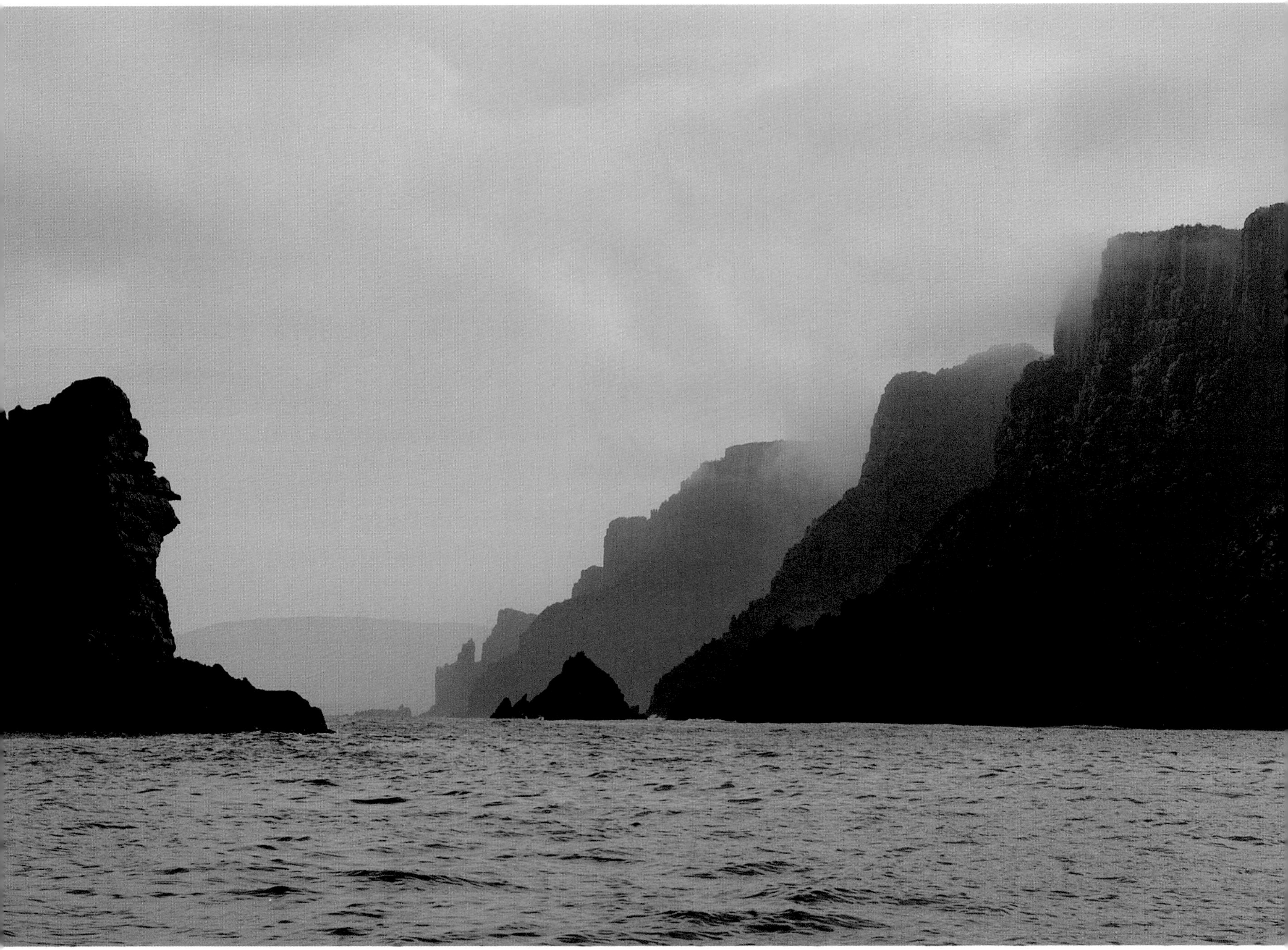

Cape Pillar, from Tasman Island, Tasmania

understand but it is most likely due to subtle changes in the ocean currents. Kelp needs cold, nutrient-rich water to survive, as it has along the coasts of California and British Columbia in Canada.

The growth rates of these plants are quite incredible. Some kelp plants are annuals, like bull kelp, which grows fast and furiously for one season and dies off in winter. Giant kelp, however, is the champion grower. They are perennials, living for a few years, during which they reach extraordinary sizes. They manage about thirty metres of growth in a year, sometimes extending by sixty centimetres in a single day. Kelp plants are anchored in shallow water, in good reach of sunlight. When a fledgeling kelp plant finds a space on a rock, it sends out small entwined roots to form a 'holdfast'. Even on fully grown giant kelp, this anchor is only fist-sized but its grip is tenacious.

On the dive I find the comparison to a terrestrial forest compelling. Small animals shelter around the holdfast, especially juveniles like baby crabs and tiny fish – just like the small animals and insects found around the roots of a tree. And like mosses and caterpillars, there are algae and grazers on the kelp too. The prettiest of these are nudibranchs – delicate, multi-coloured sea slugs with frill-like gills that sway gently to the rhythm of the sea.

Streamlined fish move easily in the water just like gannets and gulls glide on air currents, while fish on the seabed resemble animals on the forest floor. As I watch an impossibly red velvet fish towards the end of the dive, I can only think of the plumage of a bird of paradise as a comparison. In the lull between rejuvenating winter storms, visits to these magnificent sea forests lift the soul.

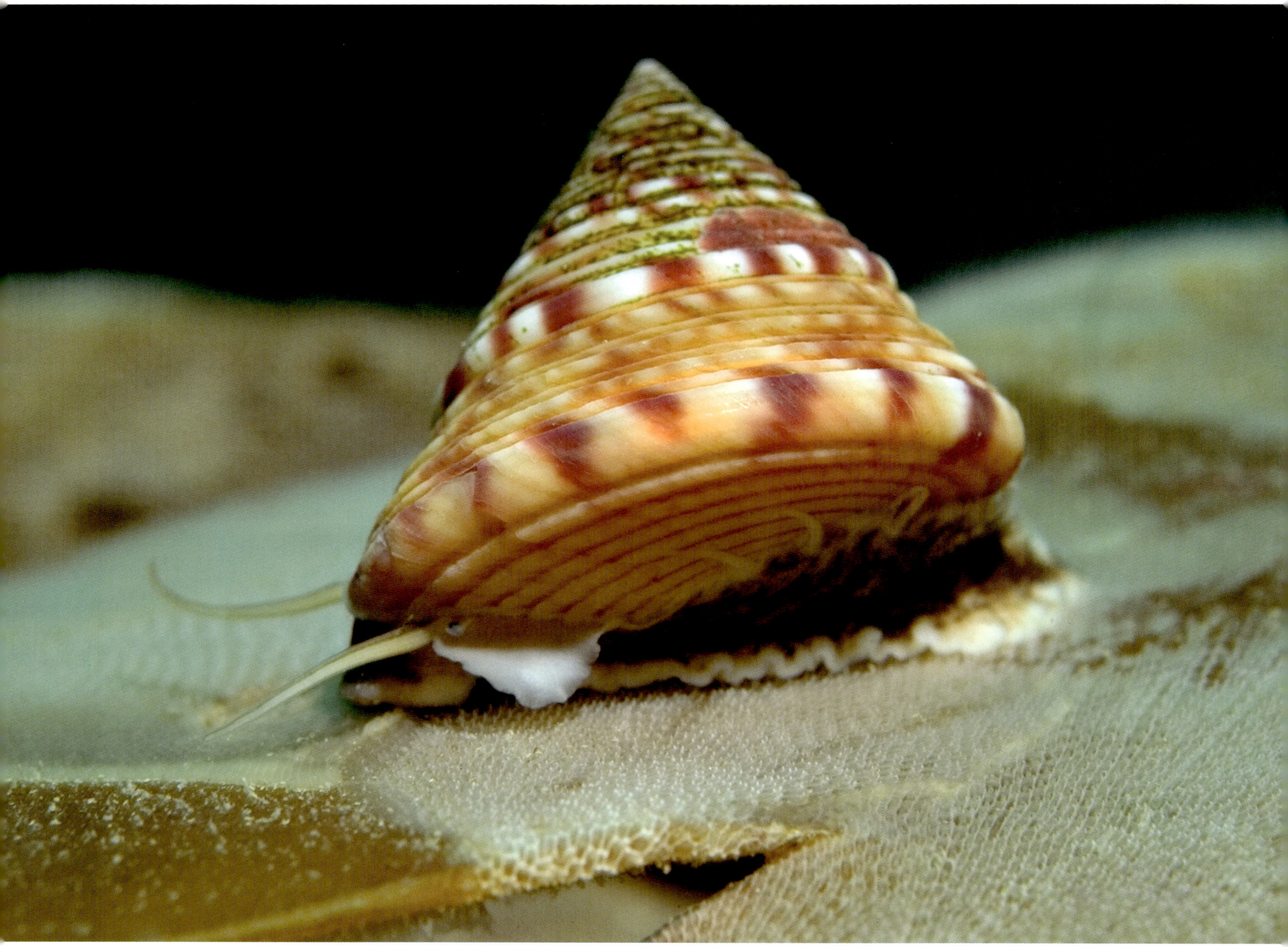

Painted Topshell, Old Head of Kinsale, Co. Cork

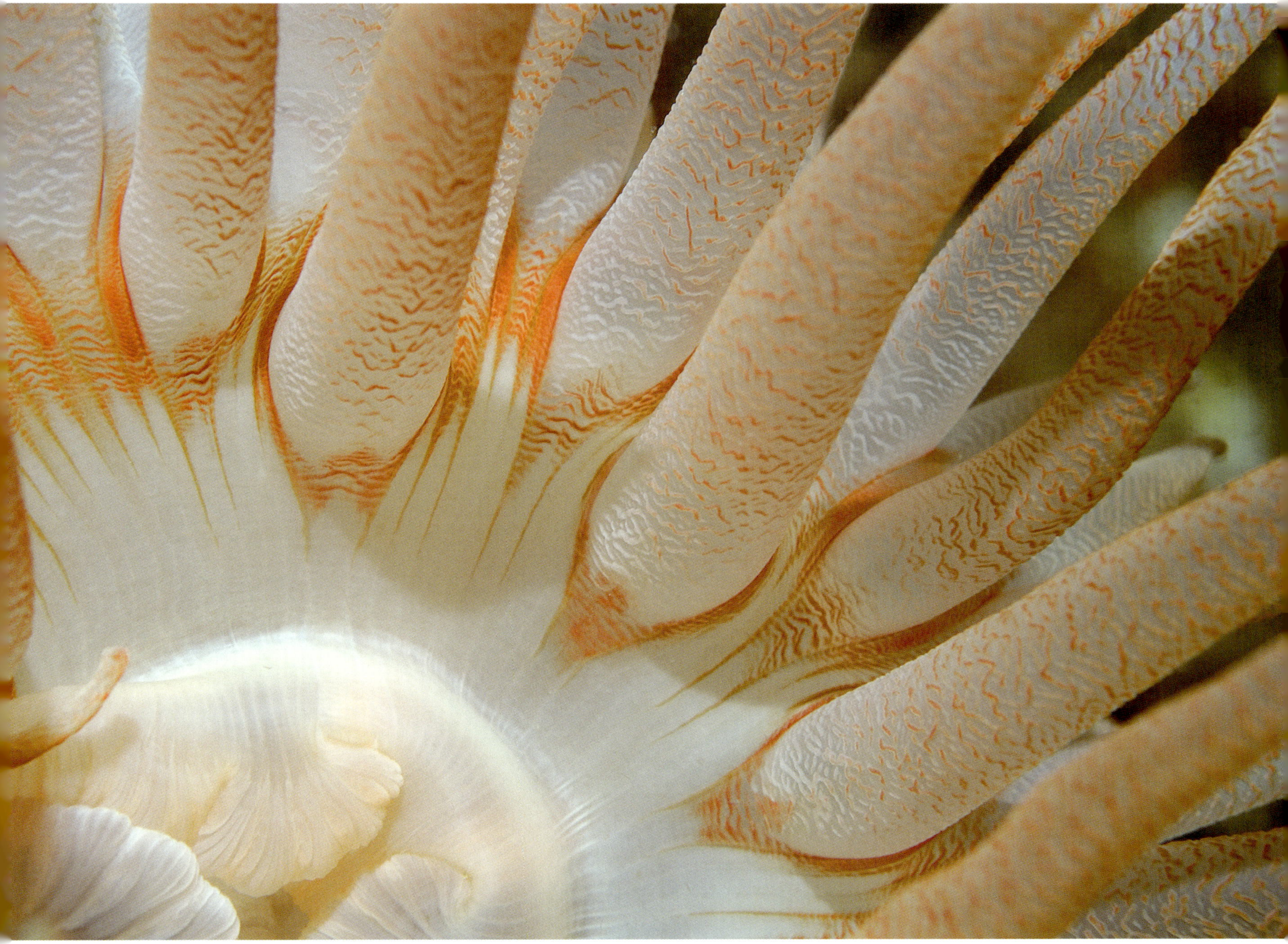

Anemone Detail, Quatsino Sound, Vancouver

… in a few minutes I was on the ladder, lowering myself into the cold water while the weight of the helmet flattened the goose bumps on my shoulders.

But once the helmet was completely submerged, it rested more comfortably. When I reached the bottom rung of the ladder I could see under the ship's hull and I glanced around. I was over a thick bed of kelp. The ends of the long strands of orange-brown seaweed reached up through the murky green water just to my feet, as if I were standing on the tops of sinuous trees in this strange underwater forest. Every now and then, the kelp tops bent in unison like long grass in an open field when a breeze passes. I could feel the shift in the water current too. It was like a cold draft and I wished I had worn more than just my bathing suit, for some clothing, though it be soaking wet, is considerable protection against the winds of the underwater world.

I grabbed the rope hanging beside the ladder and let myself slide down it into the heart of the kelp forest…

Eugenie Clark, *Lady with a Spear*, 1951.

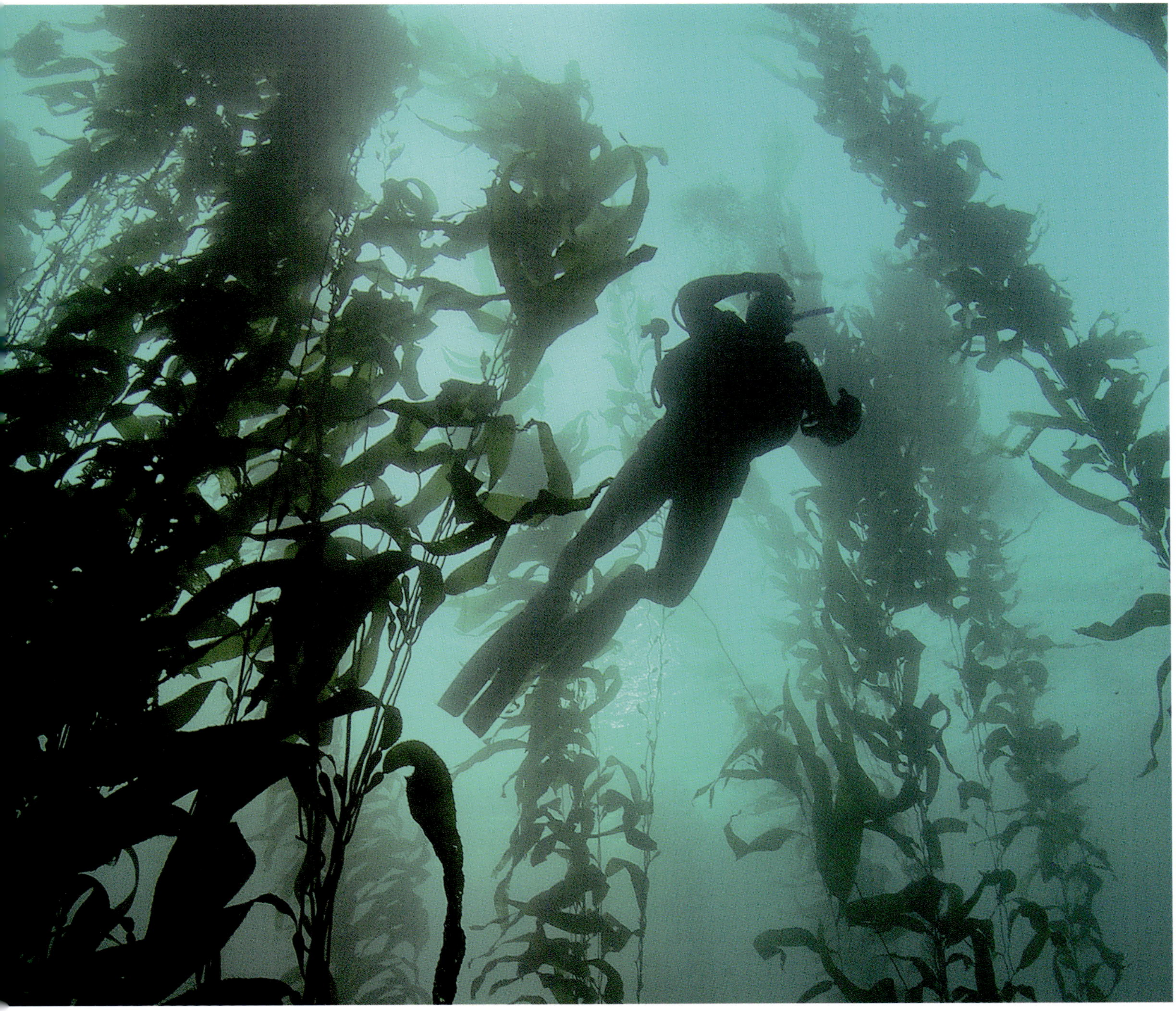

Giant Kelp, Fortescue Bay, Tasmania

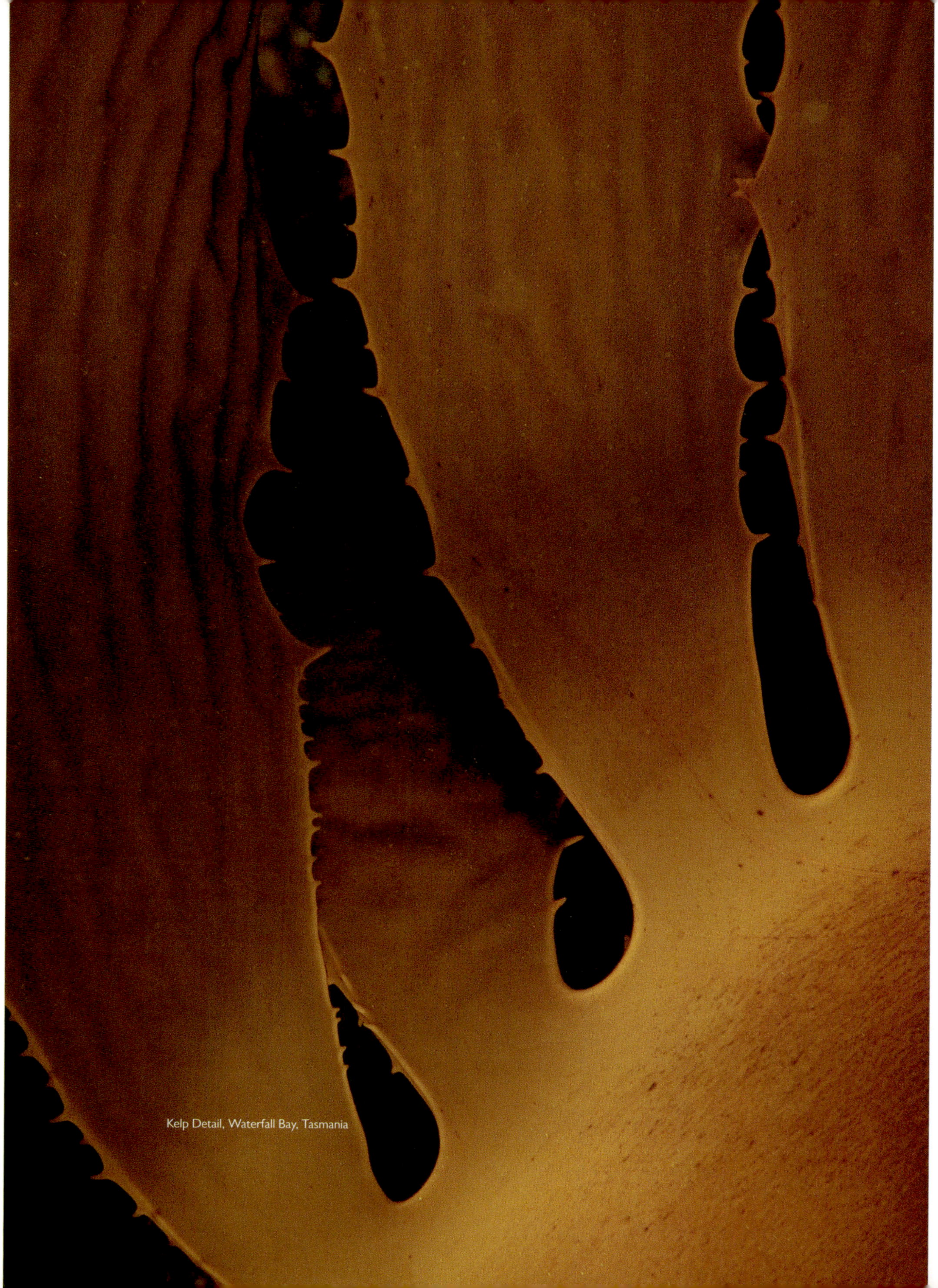

Kelp Detail, Waterfall Bay, Tasmania

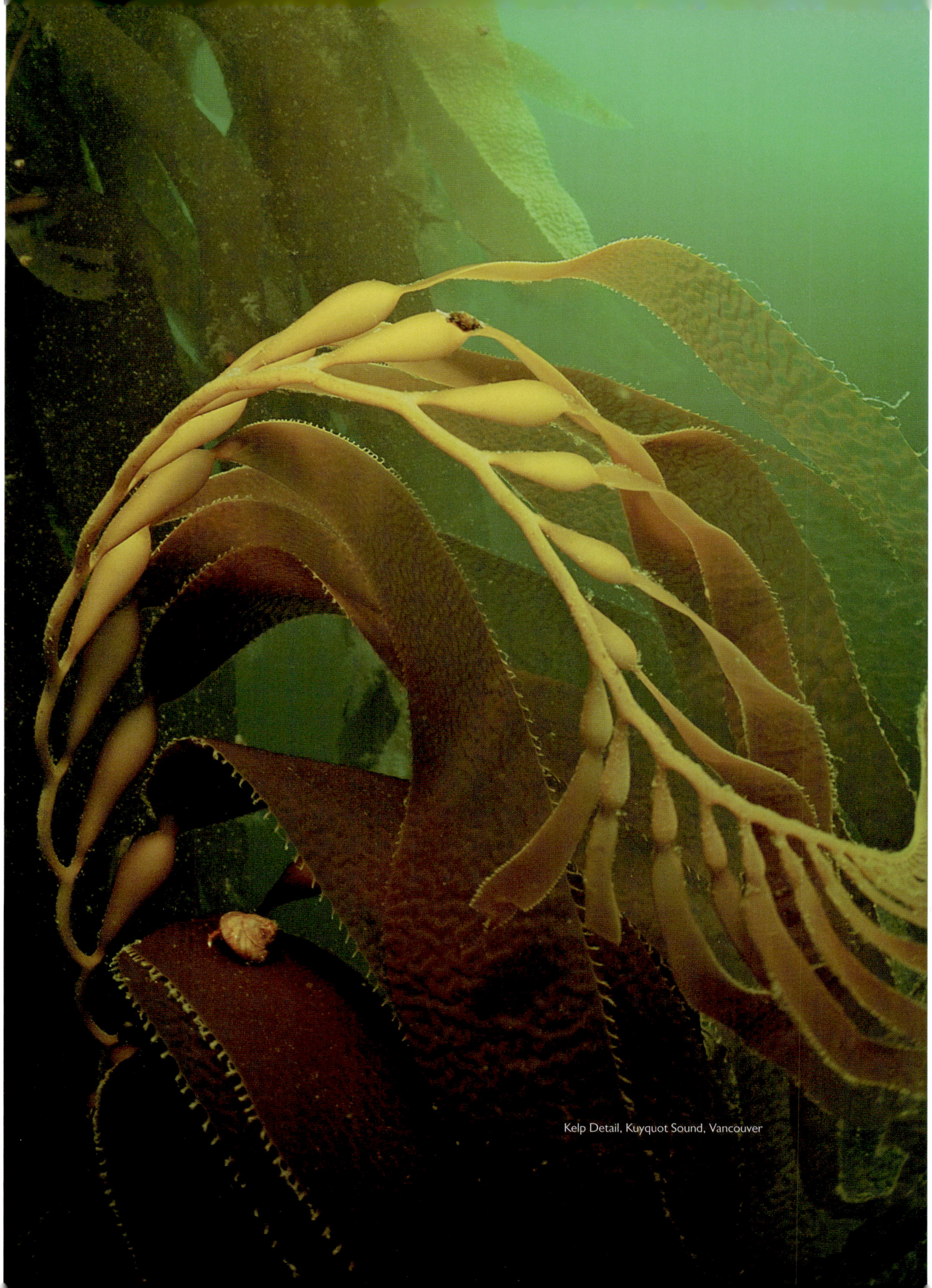

Kelp Detail, Kuyquot Sound, Vancouver

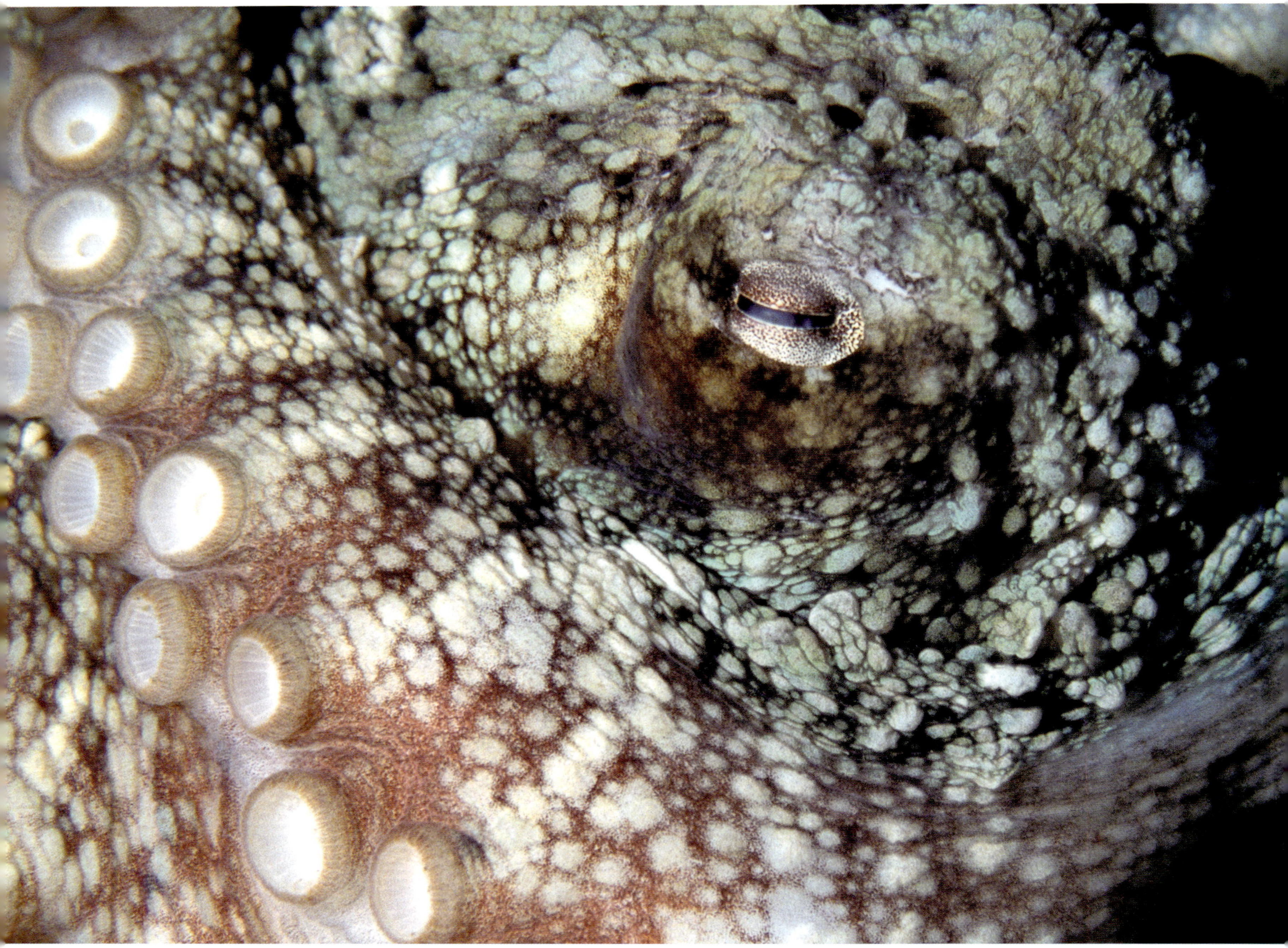

Octopus, Lanzarote, Canary Islands

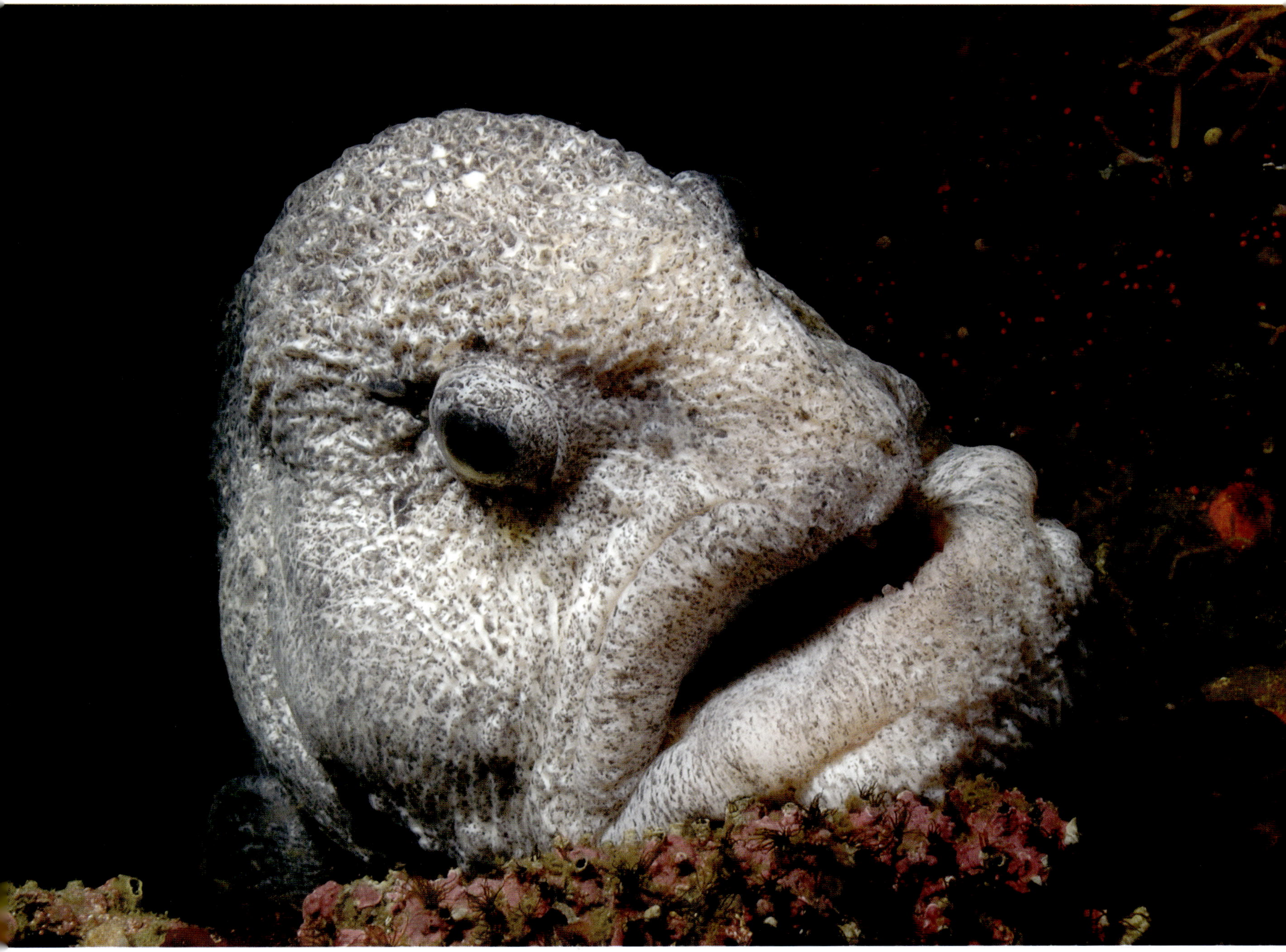

Wolf Eel, Dillon Rock, Vancouver

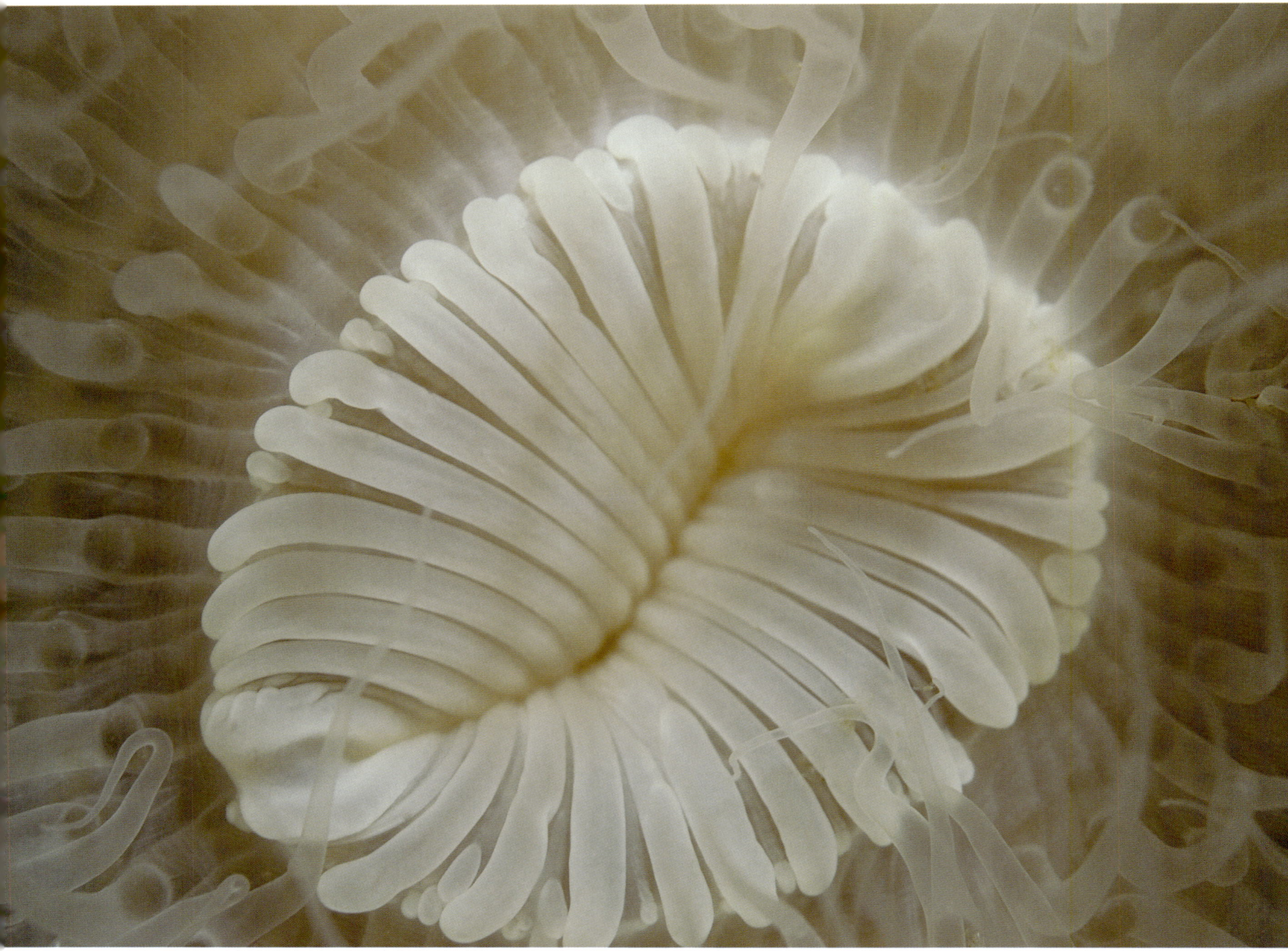

Anemone Detail, Browning Pass, Vancouver

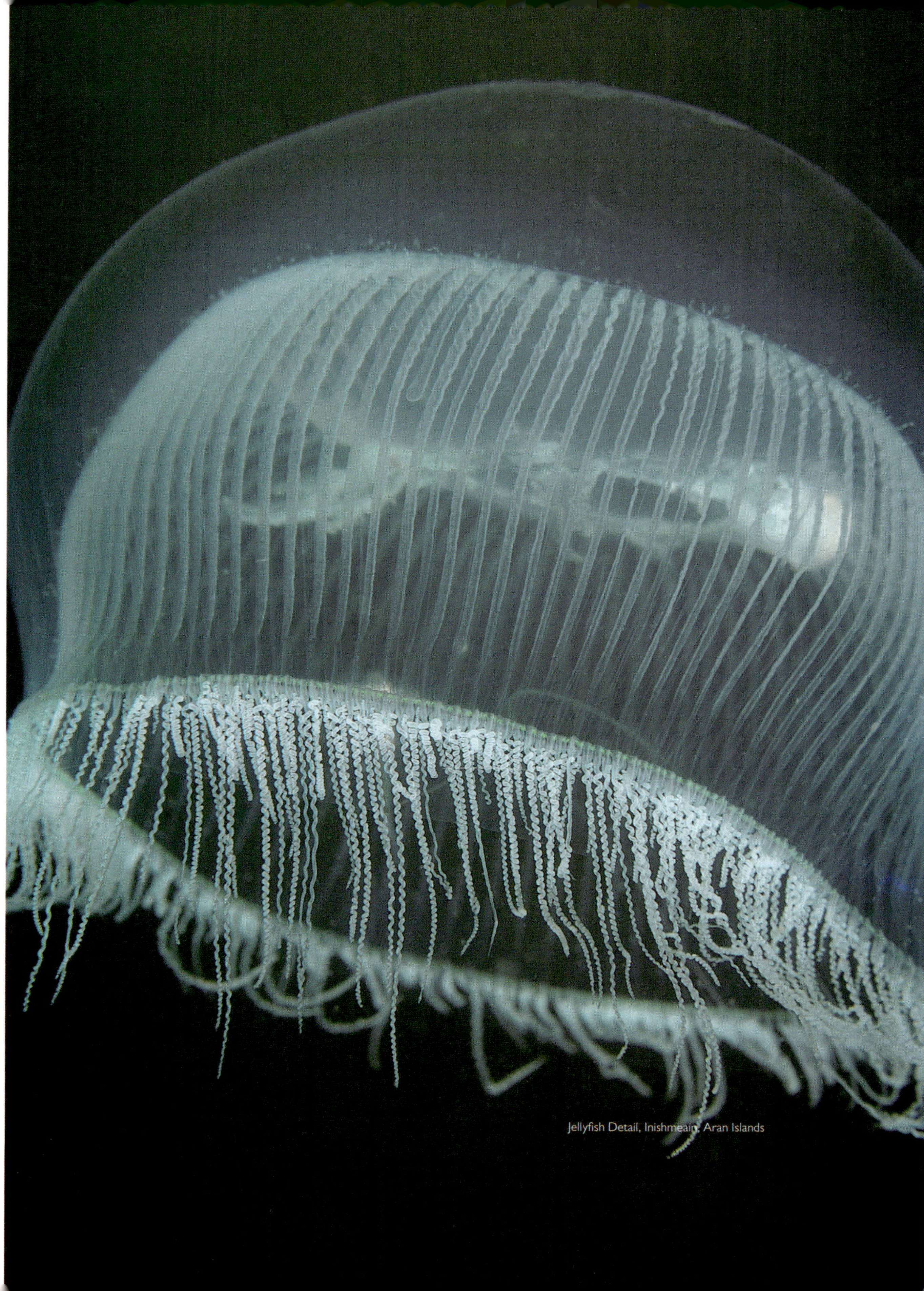

Jellyfish Detail, Inishmeain, Aran Islands

Weedy Sea Dragon, Waterfall Bay, Tasmania

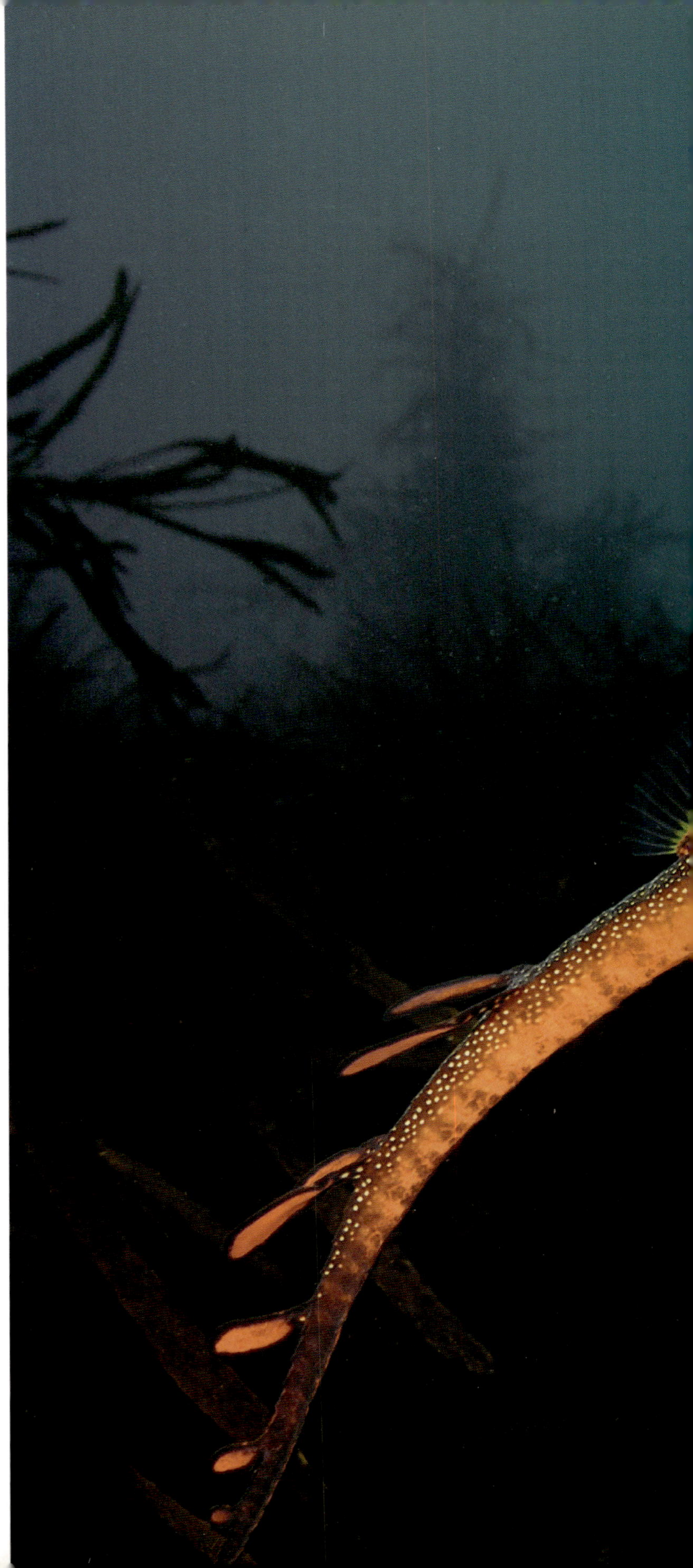

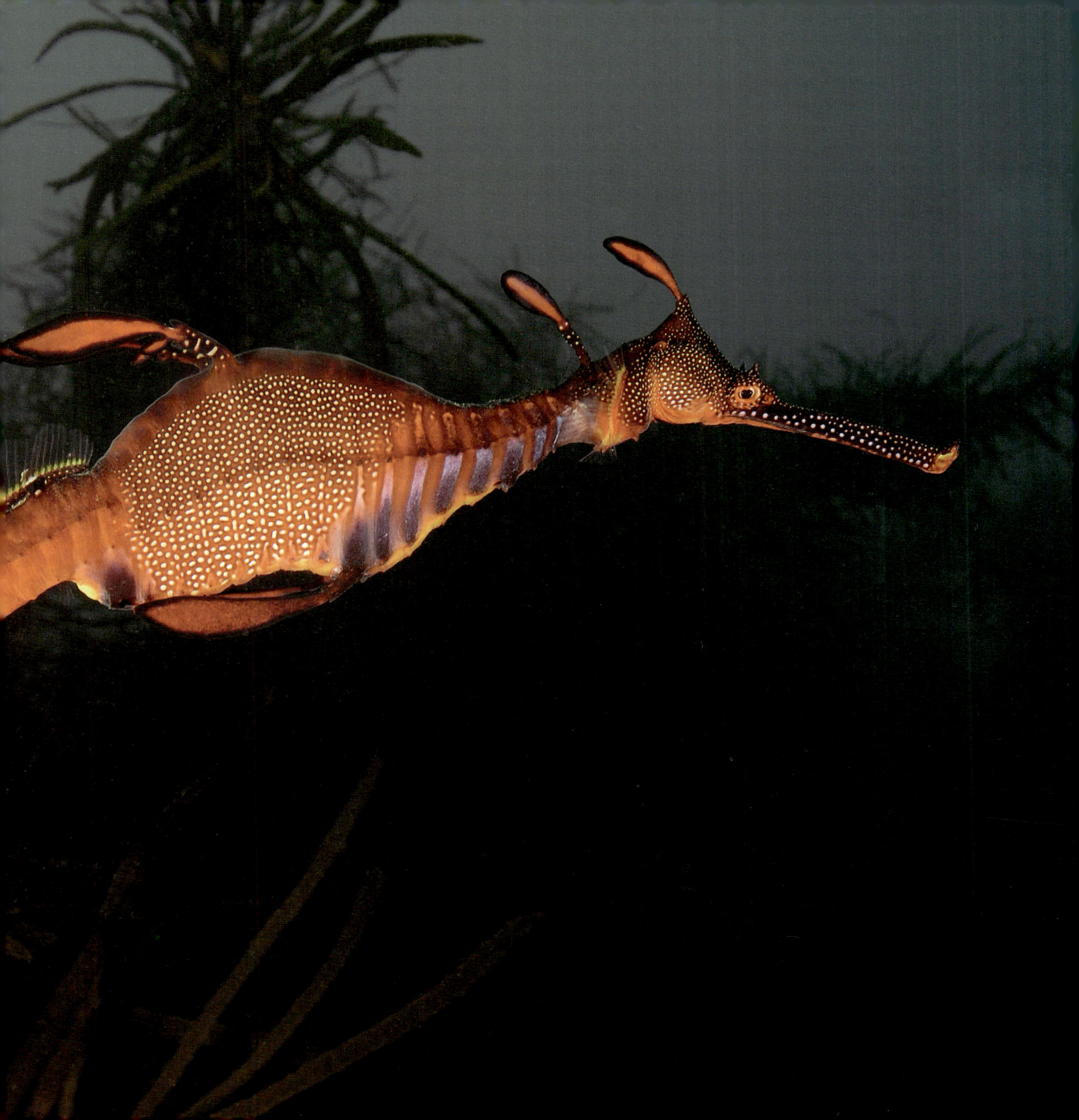

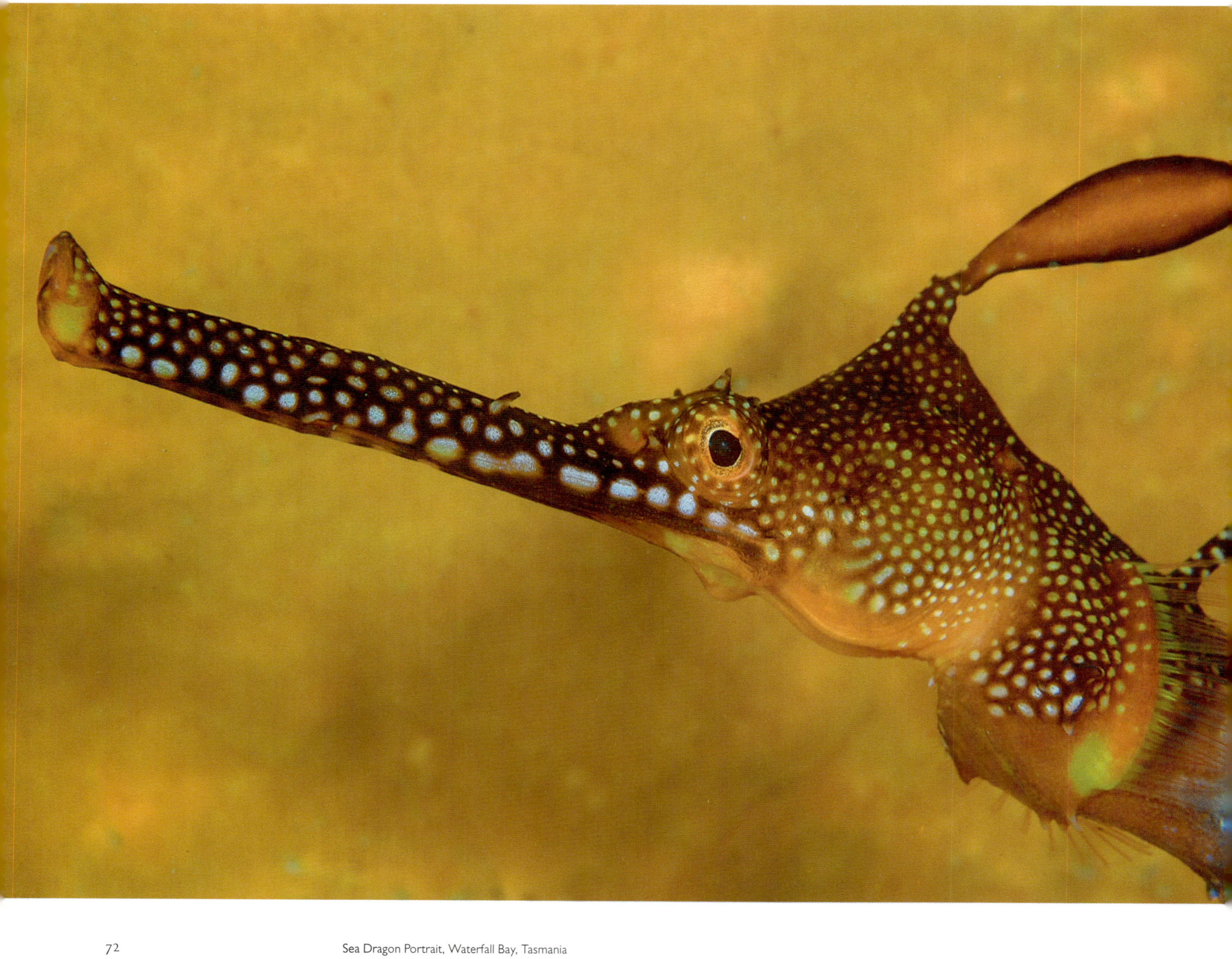

Sea Dragon Portrait, Waterfall Bay, Tasmania

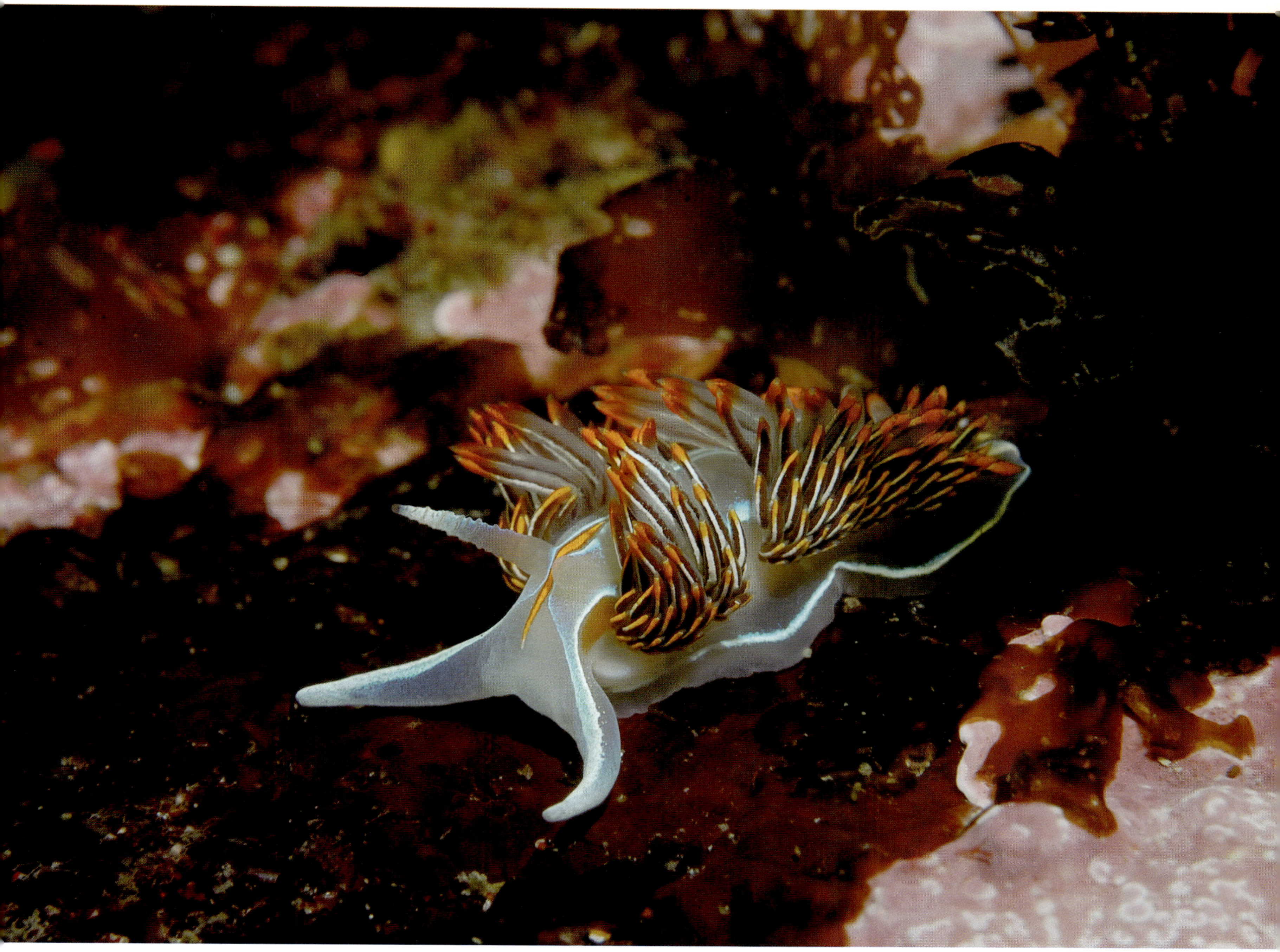

Opalescent Nudibranch, Hayden Pass, Vancouver

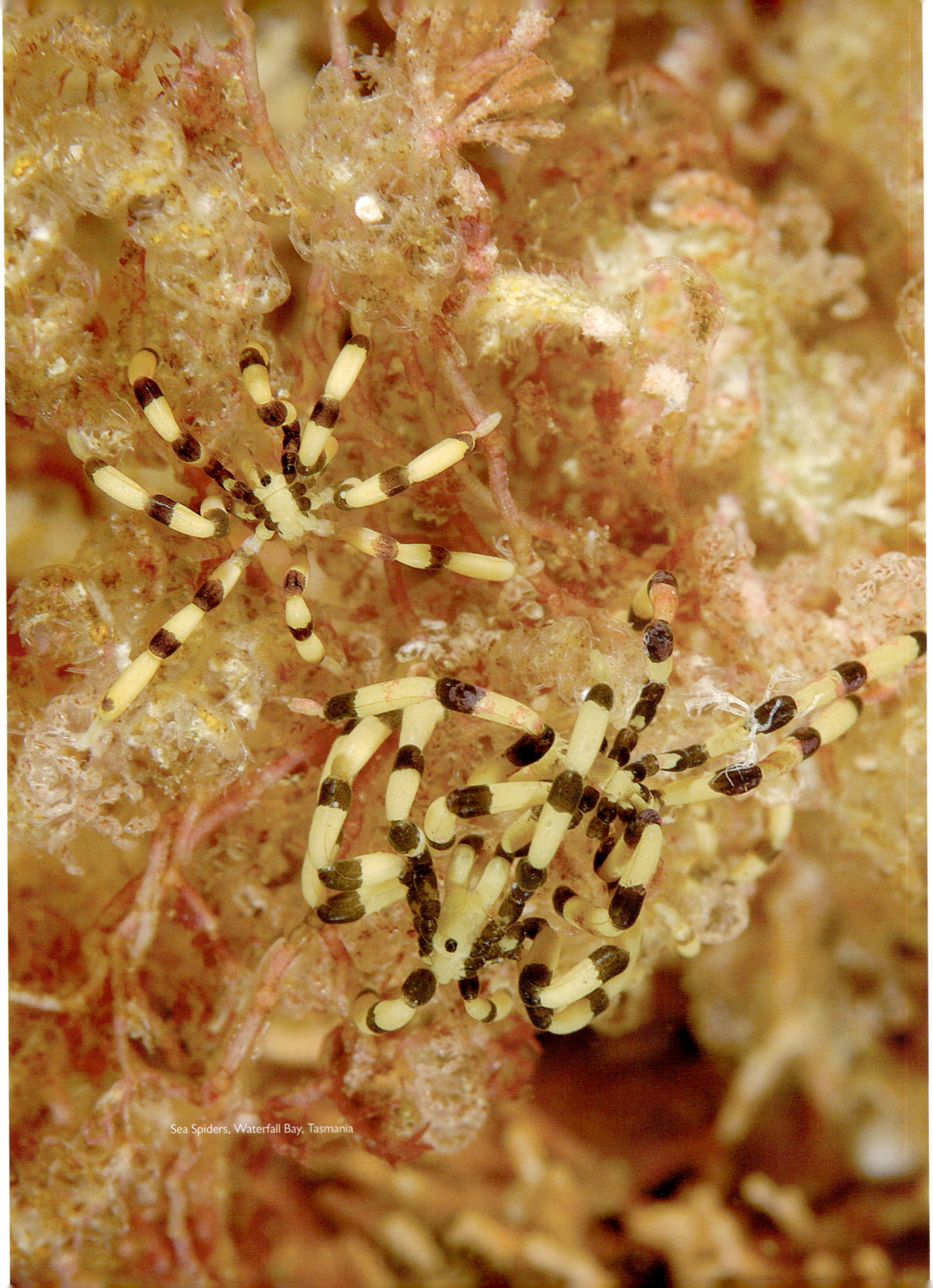

Sea Spiders, Waterfall Bay, Tasmania

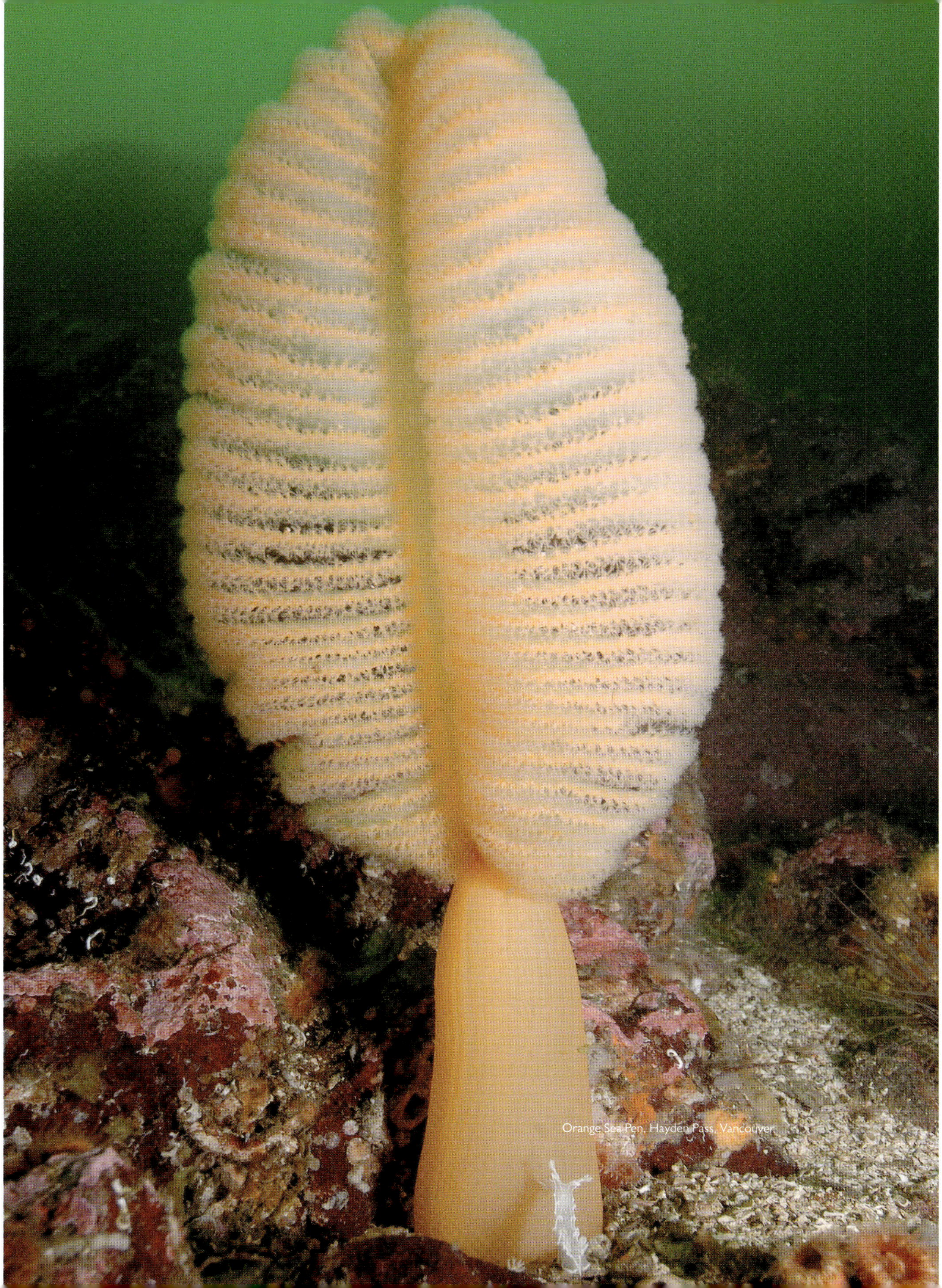

Orange Sea Pen, Hayden Pass, Vancouver

I really don't know why it is that all of us are so committed to the sea, except I think it is because in addition to the fact that the sea changes and the light changes, and ships change, it is because we all came from the sea. And it is an interesting biological fact that all of us have, in our veins the exact same percentage of salt in our blood that exists in the ocean, and, therefore, we have salt in our blood, in our sweat, in our tears. We are tied to the ocean. And when we go back to the sea, whether it is to sail or to watch it we are going back from whence we came.

John F. Kennedy, September, 1962

(Remarks of the President on the Occasion of the Dinner Given by His Excellency the Hon. Sir Howard Beale, Ambassador of Australia to the United States President John F. Kennedy, Newport, Rhode Island, September 14, 1962)

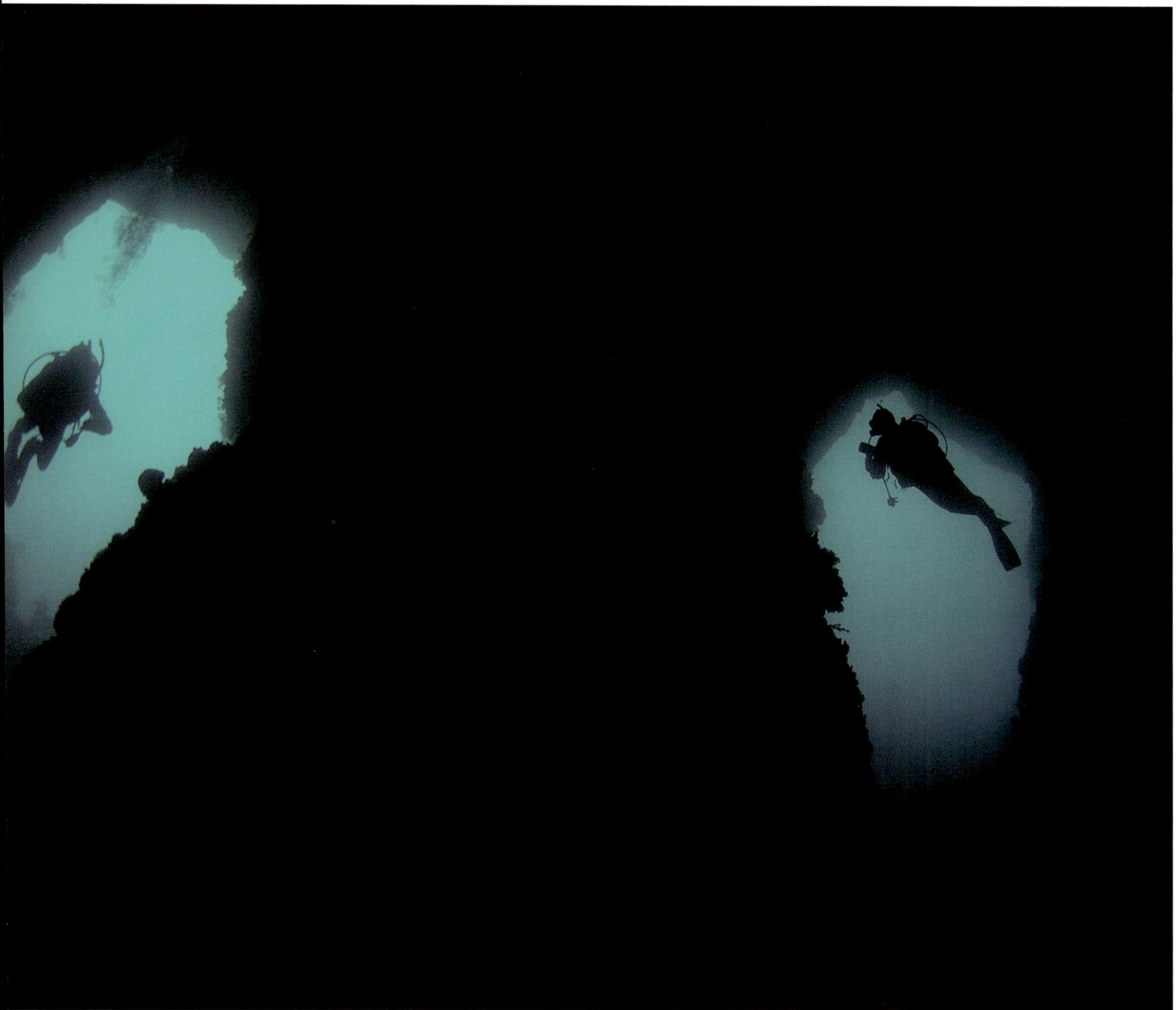

Cathedral Cave, Waterfall Bay, Tasmania

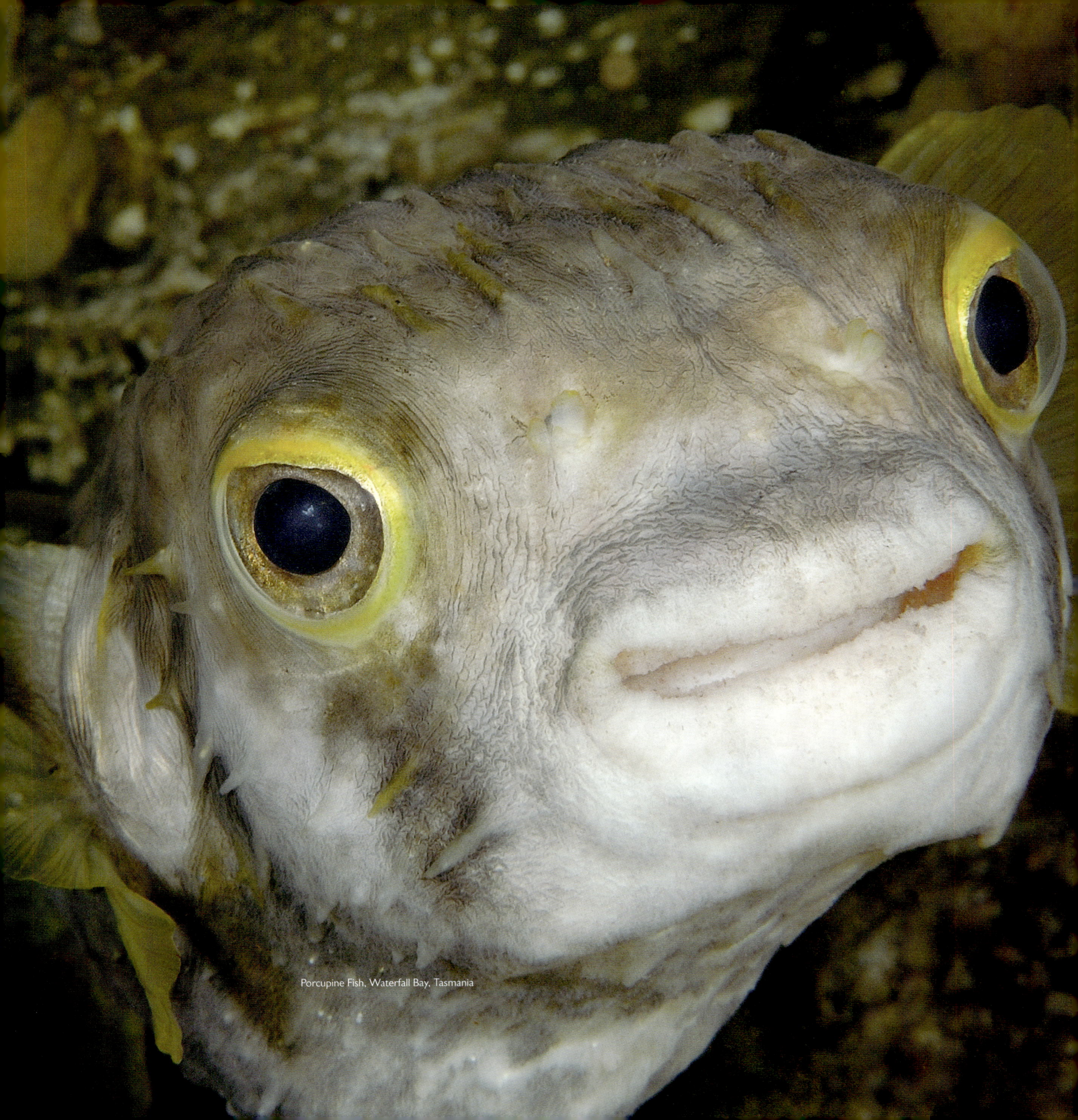

Porcupine Fish, Waterfall Bay, Tasmania

FACE TO FACE

In the heavy green depths of Dingle harbour, a sleek silhouette suddenly appears and sonar squeaks fill my ears. Then as quickly, it is gone. We are giddy with excitement – it is true – a friendly dolphin has taken up residence and can be seen regularly at the mouth of this busy Kerry harbour. It is the weekend and the word has spread. Canoeists, divers, snorkelers, eager children on freshly painted fishing boats – even jet-skiers, all hope for a glimpse of this four metre, bottlenose dolphin named 'Fungi' by the local fishermen. They were among the first to notice his regular appearances, playing in the bow waves of their boats, as they returned from a day's fishing. A dolphin industry has mushroomed, as people flock to see his playful antics. He shares his time equally among his visitors, moving from diver to snorkeler to boat.

Early next morning, most of the visitors have gone and the boats have not yet left the harbour. We snorkel out from a small beach near the harbour mouth and wonder if we need to attract his attention. With echo-location that can detect a fish from 100 metres I reason that he is aware of our presence long before we see him. Looking around at the surface, there is no tell-tale dorsal fin or puff from a blowhole. I dip my face back in the water and I am startled to be nose to nose with him. The permanent smile is contagious – I laugh out loud with child-like excitement. A couple of hours later, we are shivering, exhausted and long out of film as we make our way back to the beach, high with pure joy. Playing with a wild dolphin is an experience that stays with you. It is a kinship with the sea itself, a contact with an ambassador that could well have been sent from a higher being.

The big animals, just like those on an African safari, are the most exciting to encounter at sea. Seeing a fifty ton whale and watching its flukes gracefully disappear below the surface, is a humbling experience. Watching the ease with which a seal circles round and nibbles on your fins, makes you realise just how cumbersome we are in the water. And seeing the apex predators stalk the cool waters of the south Atlantic is simply awe inspiring.

The Great White Shark is the ocean's most fearsome predator. It is widely distributed in all seas, but particularly in the cool waters of South Africa and Australia. They are solitary animals and we know little about them. Getting close to them is difficult, as they are essentially man-avoiders. Food, or the smell of it, is the only way we can deliberately attract these animals into close quarters, in order to observe them. During the southern winter, near Dyer Island in South Africa, white sharks are regularly seen patrolling the waters. On nearby Geyser Rock, thousands of seal pups have just been weaned and are venturing into the sea, ignorant of the sharks. The turbid water gives excellent cover for a sudden strike and a rich meal.

On arrival at the small town of Gansbaai, the sea is white as large waves crash ashore. Dyer Island is barely visible in the distance. We listen, with raised eyebrows, as the skipper of the ten metre catamaran tells us the day's plan. 'It will be a little rough on the way out, so hold on tight', he grins. 'Once we anchor, we will try to attract the sharks to the boat – if they stay, we launch the cage. Once you are in the water, there is only one rule: no part of you can be outside the cage at any time. You cannot see a shark come from underneath.' We would be at sea for four to five hours. We might see sharks, but we would need a little luck.

Thrashing through the water, the twin outboard engines roar in protest to the high seas. In the lee of the island, the breakers smooth out to long, steep swells. The breakers thunder into Geyser Rock, the booms mixing with the cacophony of thousands of seals barking and gulls calling from overhead. Driving rain stings exposed cheeks as hoods are drawn tight. A large plastic bait box is opened and an unforgettable stench fills the air. Two onion bags, filled with offal from a local fish factory, are strung over the sides of the boat. An oily slick trickles into the current. Any sharks crossing it, should follow the scent to investigate its source. At the same time, a heavy line with bait and a float is thrown aft – this lures the sharks close to the boat. On such a winter's day I doubt if we will see anything.

Within minutes, however, a trademark dorsal fin passes close to the bait. The shark researcher, slowly draws in the line. The shark criss-crosses a path towards us. It is huge, its shape now visible in the grey water. Closer and closer, then a lunge to swallow the bait whole. The famous jaws protruding and the eyes rolling back for protection, each bite a determined attempt to take the food. The line is held tight and the enormous head clears the water and lunges once more for the bait. The rope is pulled clear as the researcher grasps the snout and lifts it upwards, the shark momentarily confused. A final, searching bite, shows her gaping pink throat and rows of serrated teeth, water swirling around the mouth as she slips quietly back beneath the waves.

Such close encounters in the sea are rare and memorable. A one-to-one meeting with even the smallest of fish is special, their big eyes staring incredulously at the bizarre monster in front of them. And occasionally, a marine resident will relax in your presence and allow you into their space. When that intimate moment comes, our different worlds merge briefly, and we are one with our sea-dwelling neighbours.

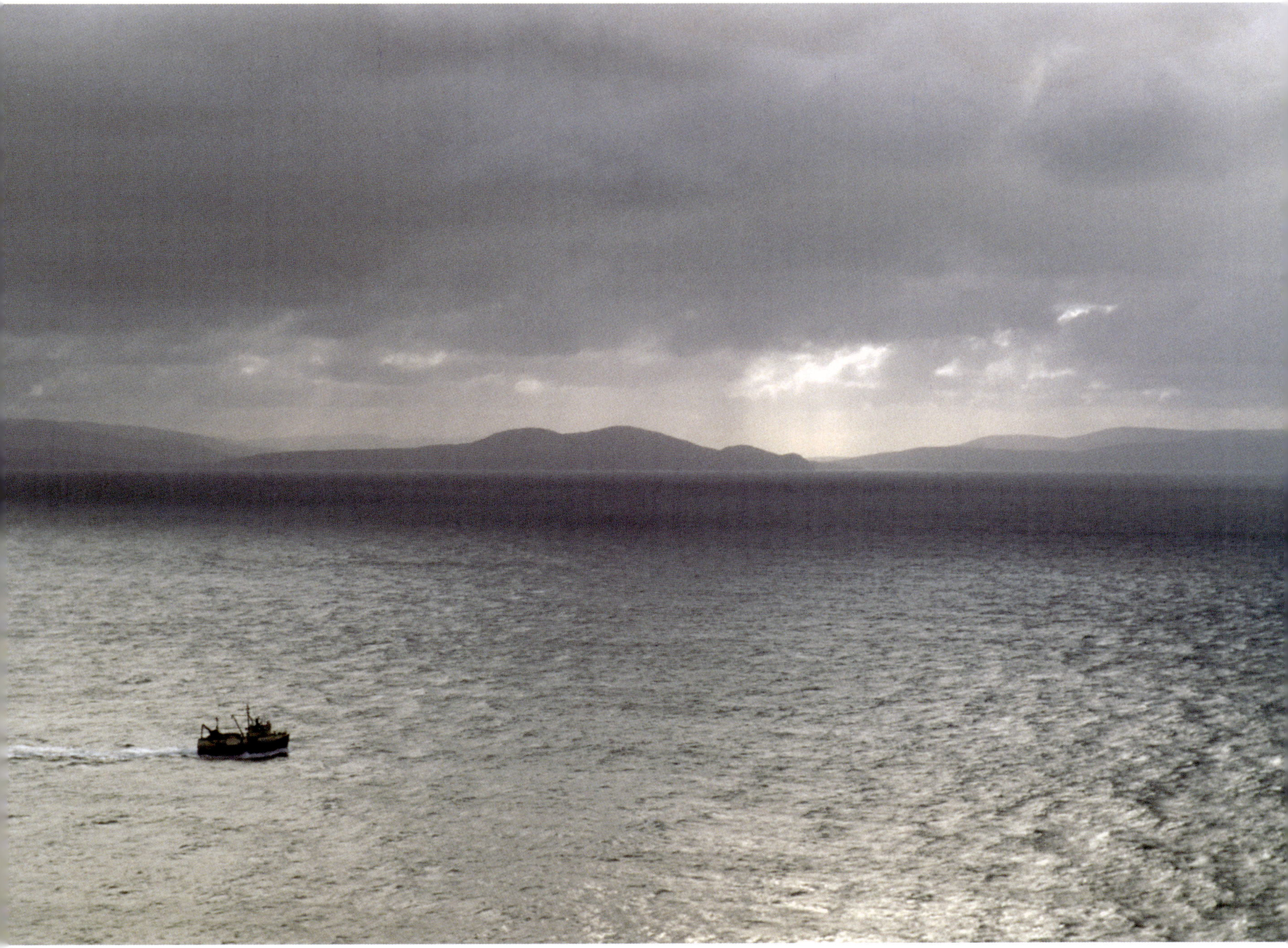

Dingle Bay, Co. Kerry

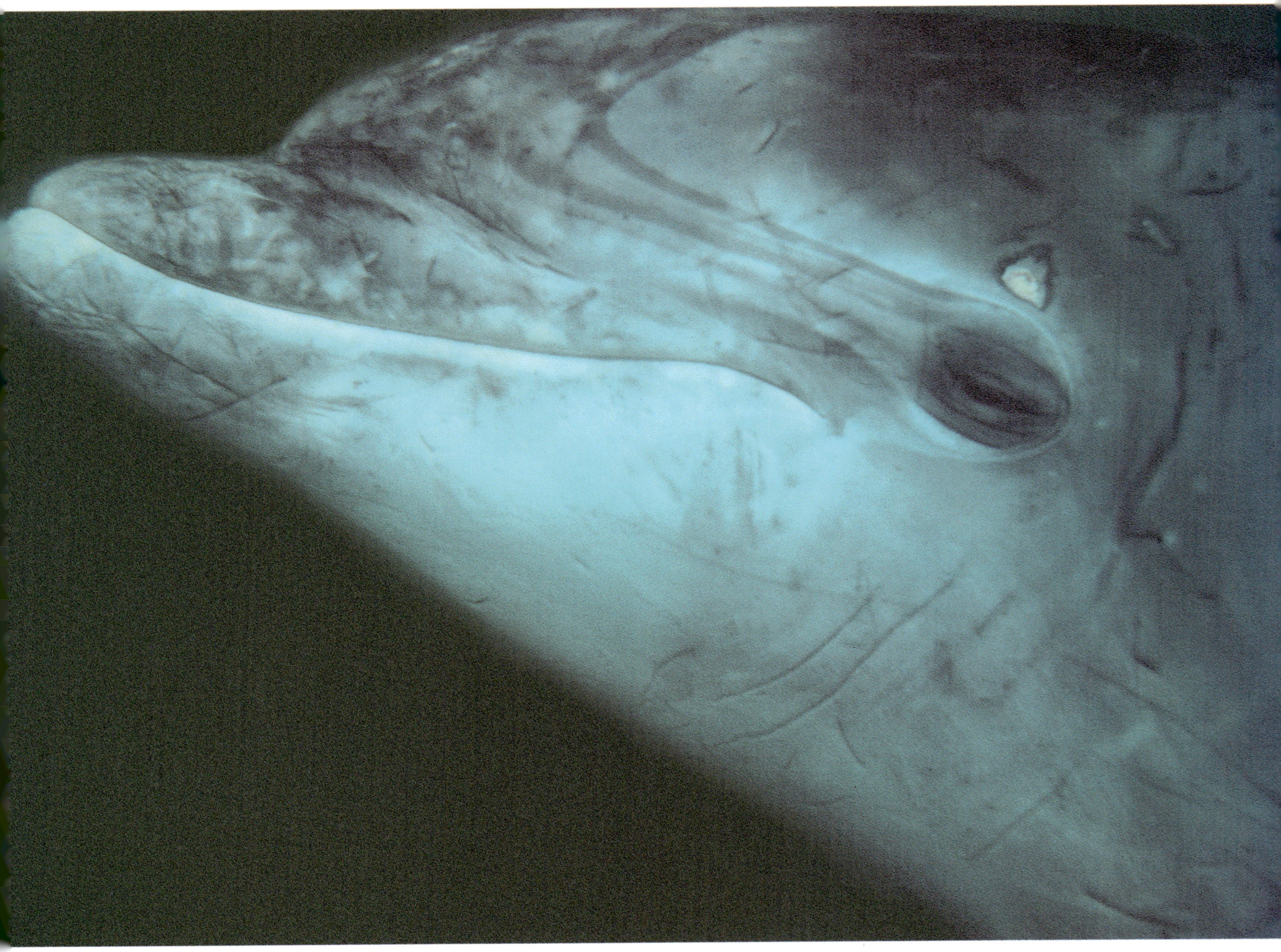

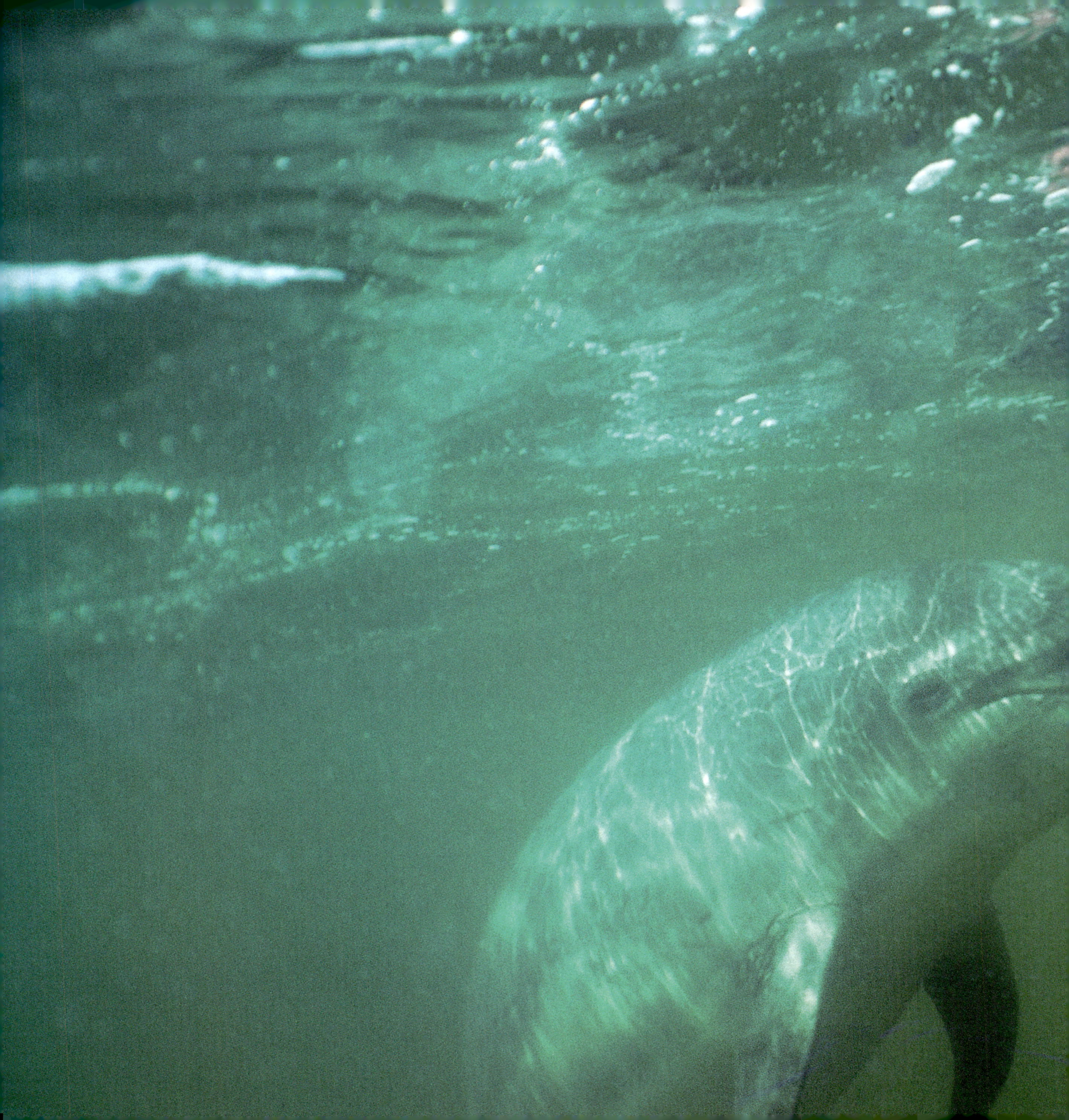

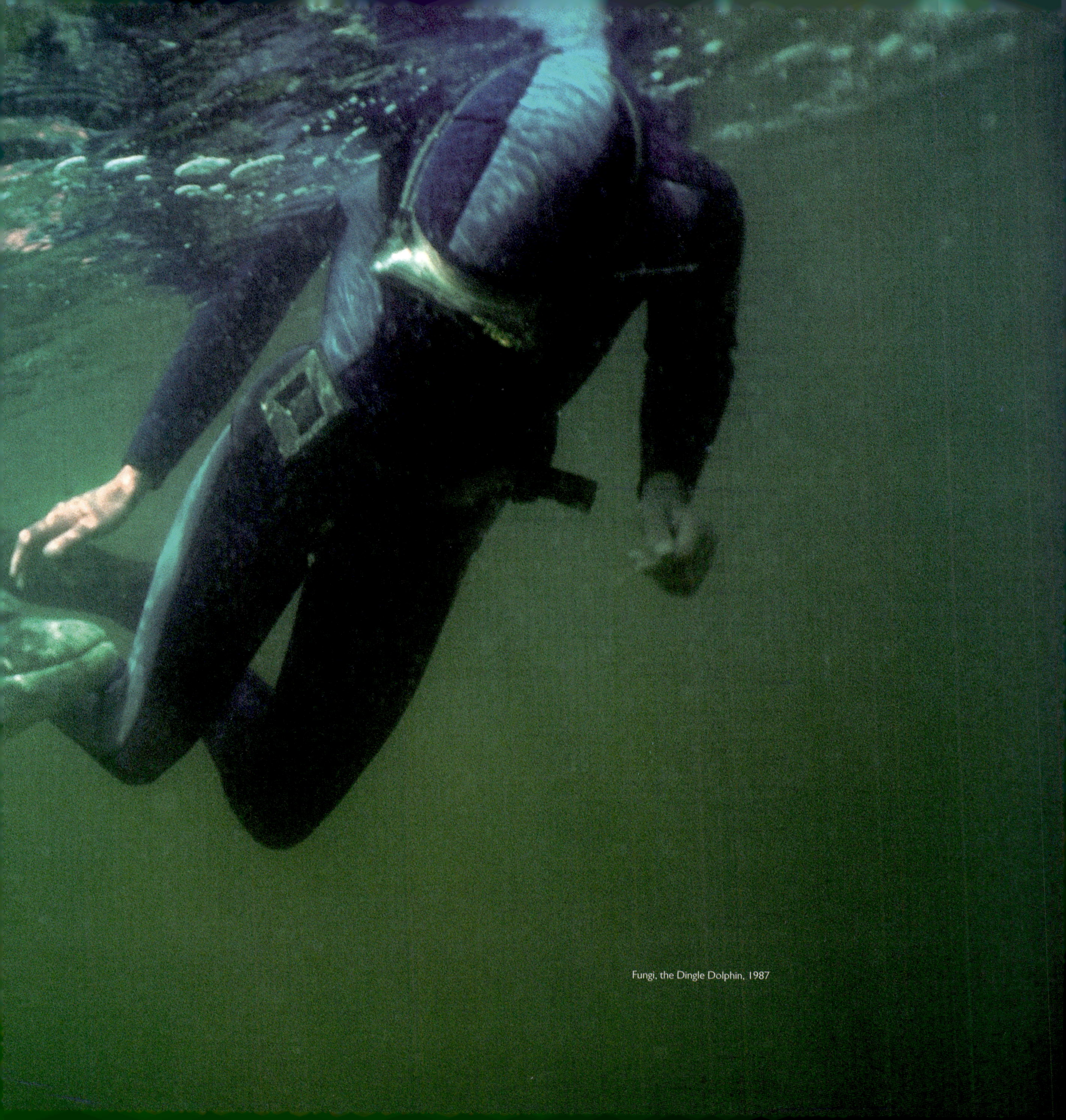

Fungi, the Dingle Dolphin, 1987

Many suspended together, forever apart, Each one alone with the waters, upon one wave with the rest.

A magnetism in the water between them only.

D.H. Lawrence, 'Fish', *Birds Beasts and flowers*

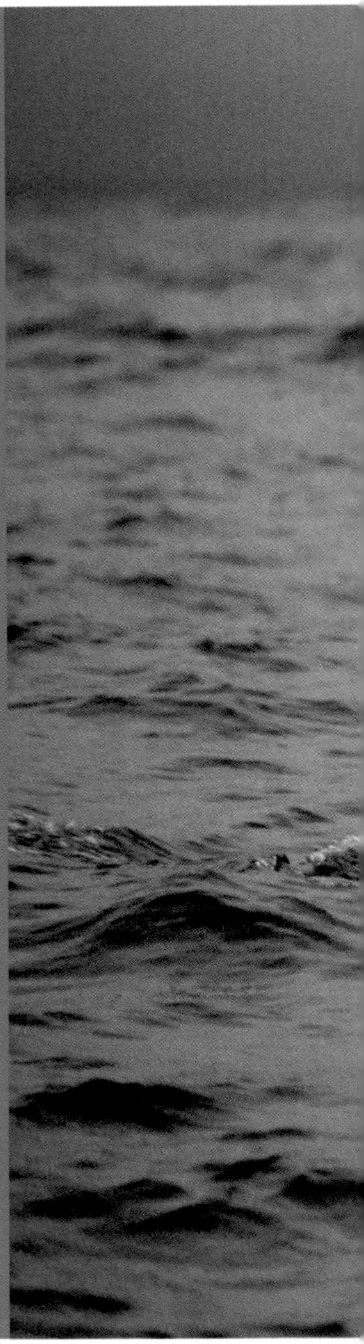

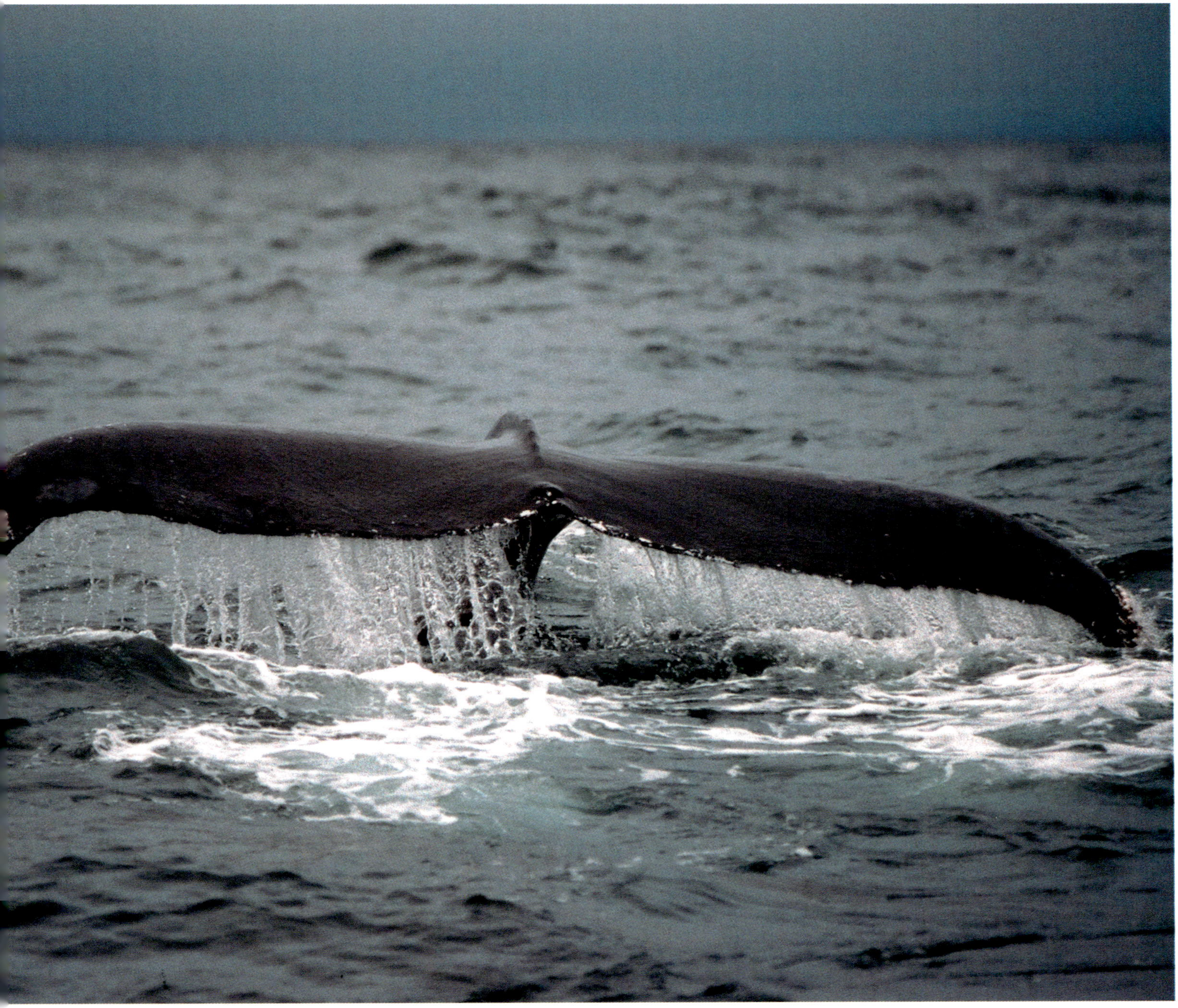

Humpback Whale, West Cork

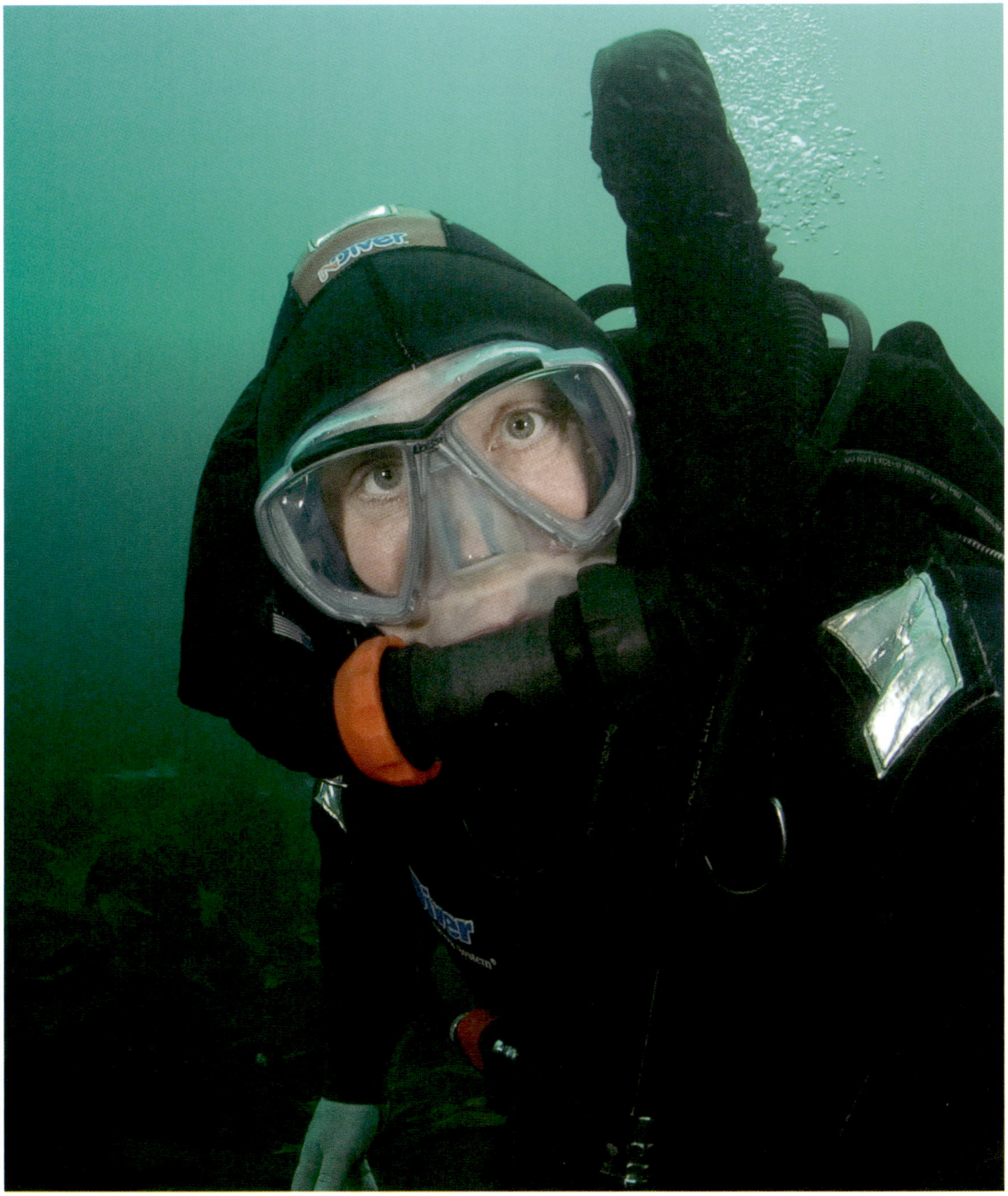

Eye to Eye…

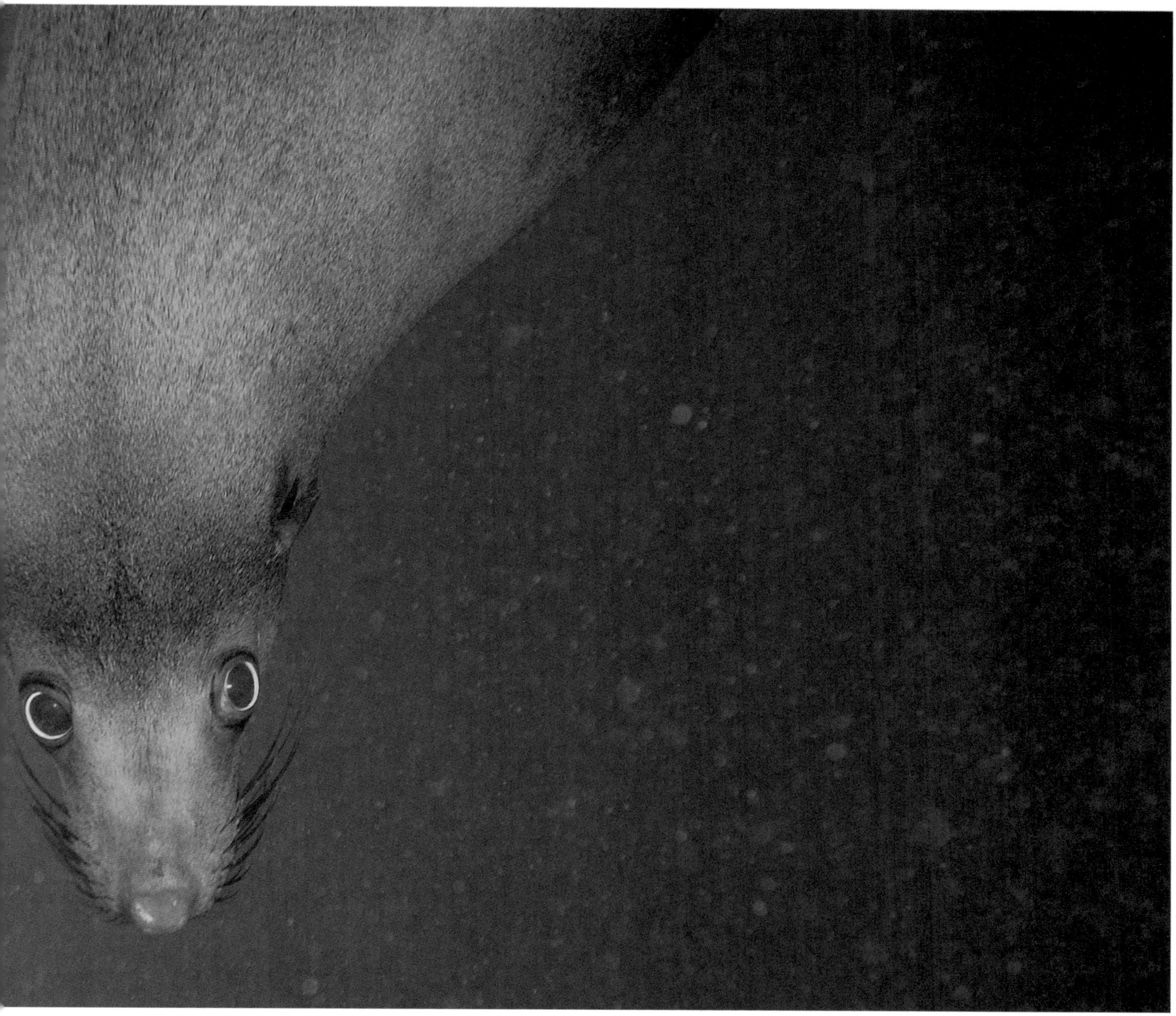

... Australian Fur Seal, Hippolyte Rocks, Tasmania

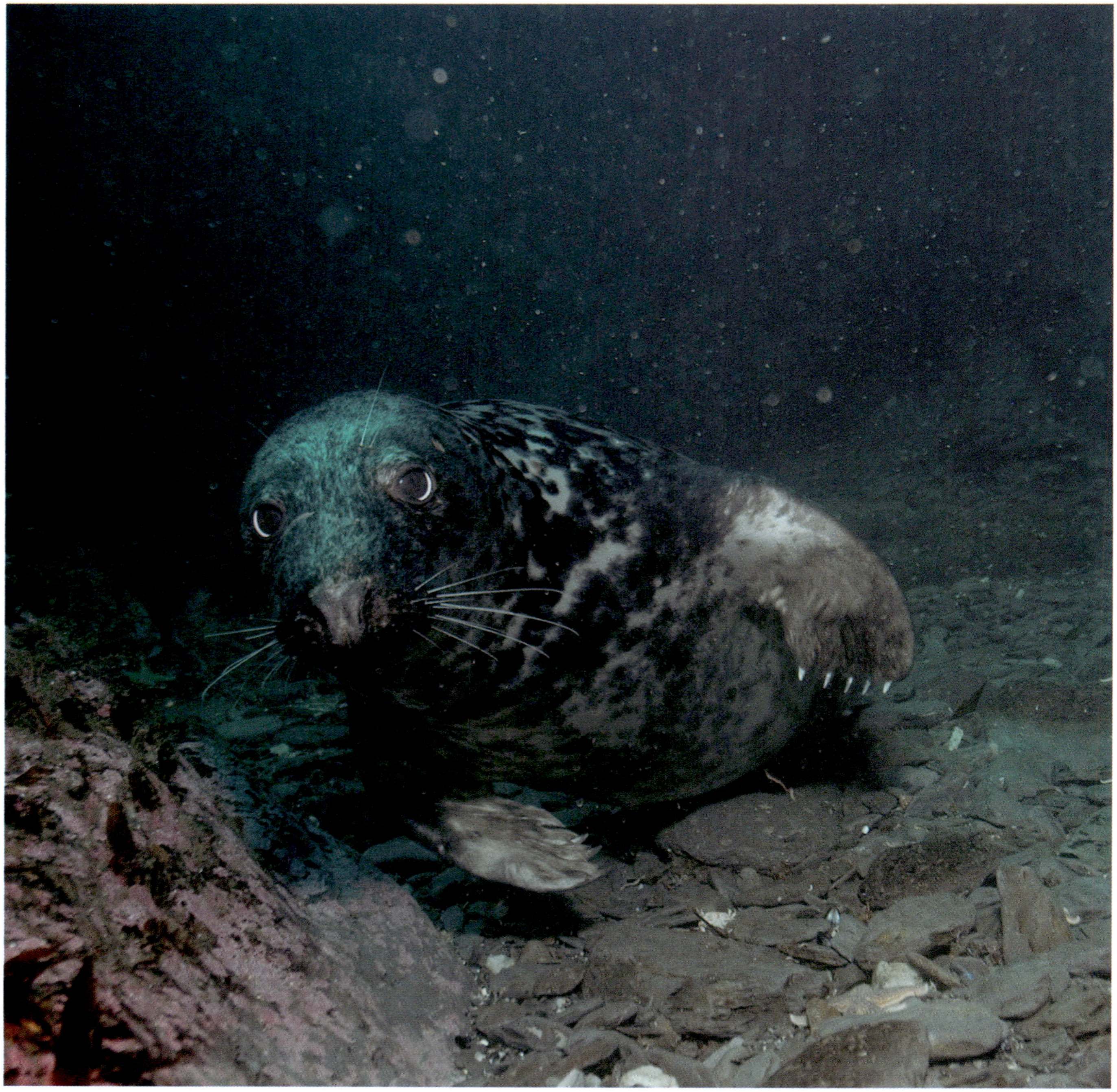

Grey Seal, Skellig Rocks, Co. Kerry

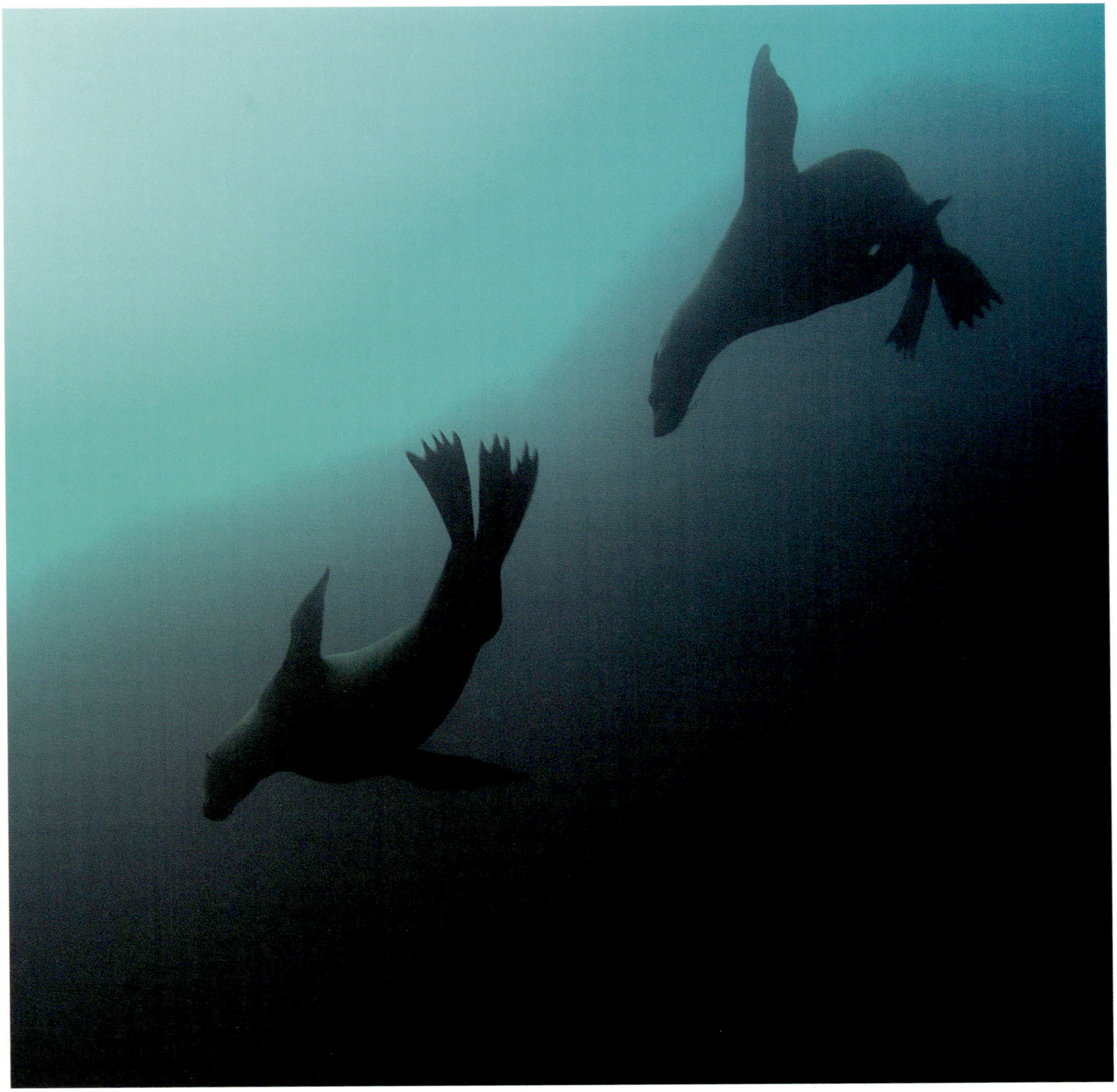

Fur Seals, Hippolyte Rocks, Tasmania

His entire form is fluid, weaving from side to side; his head moves slowly from left to right, right to left, timed to the ryhthm of his motion through the water. Only the eye is fixed, focused on me, circling within the orbit of the head, in order not to lose sight for a fraction of a second of his prey or, perhaps, or his enemy.

His skin is creased with a thousand silky furrows at every movement of his body, emphasizing each pattern of incredible muscle. The crystalline water has ceased to exist; he is there in the unbelievable purity of the void and nothing separates us any longer.

There is no threat, no movement of aggression. Only a sort of nonchalant suspicion is apparent in the movements and attitudes of the shark, and yet he generates fear. Amazed and startled, filled with apprehension, circling with movements as slow and silent as possible, I try to keep him constantly in front of me.

Philippe Cousteau, *The Shark: Splendid Savage of the Sea*, 1970

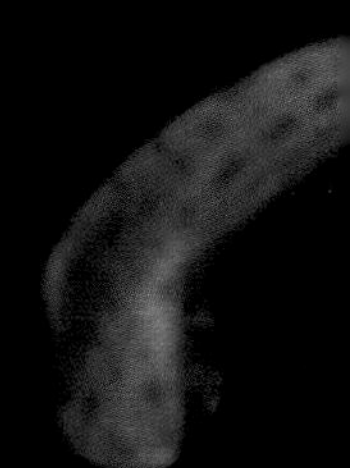

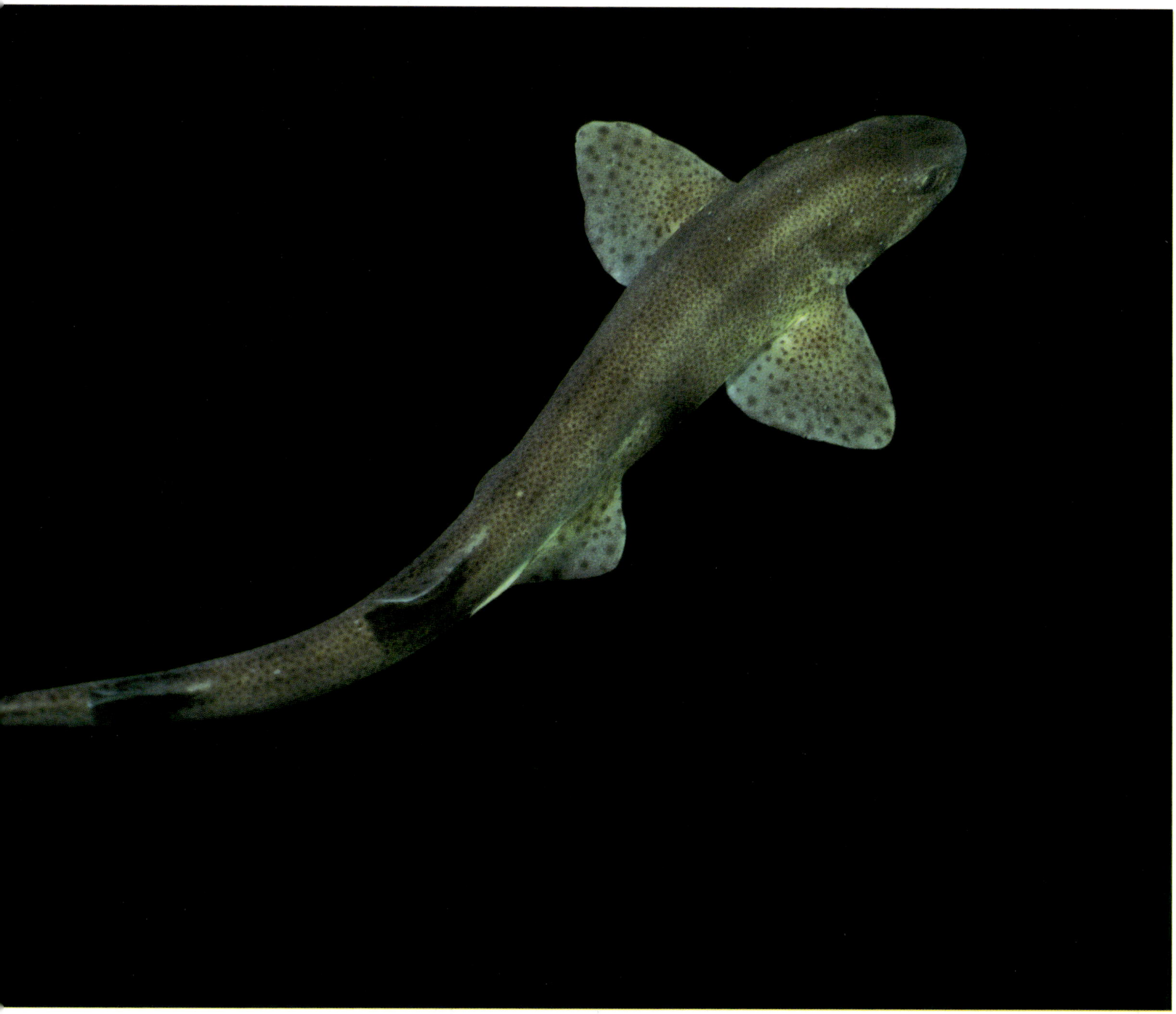

Spotted Dogfish, Inishbofin, Co. Galway

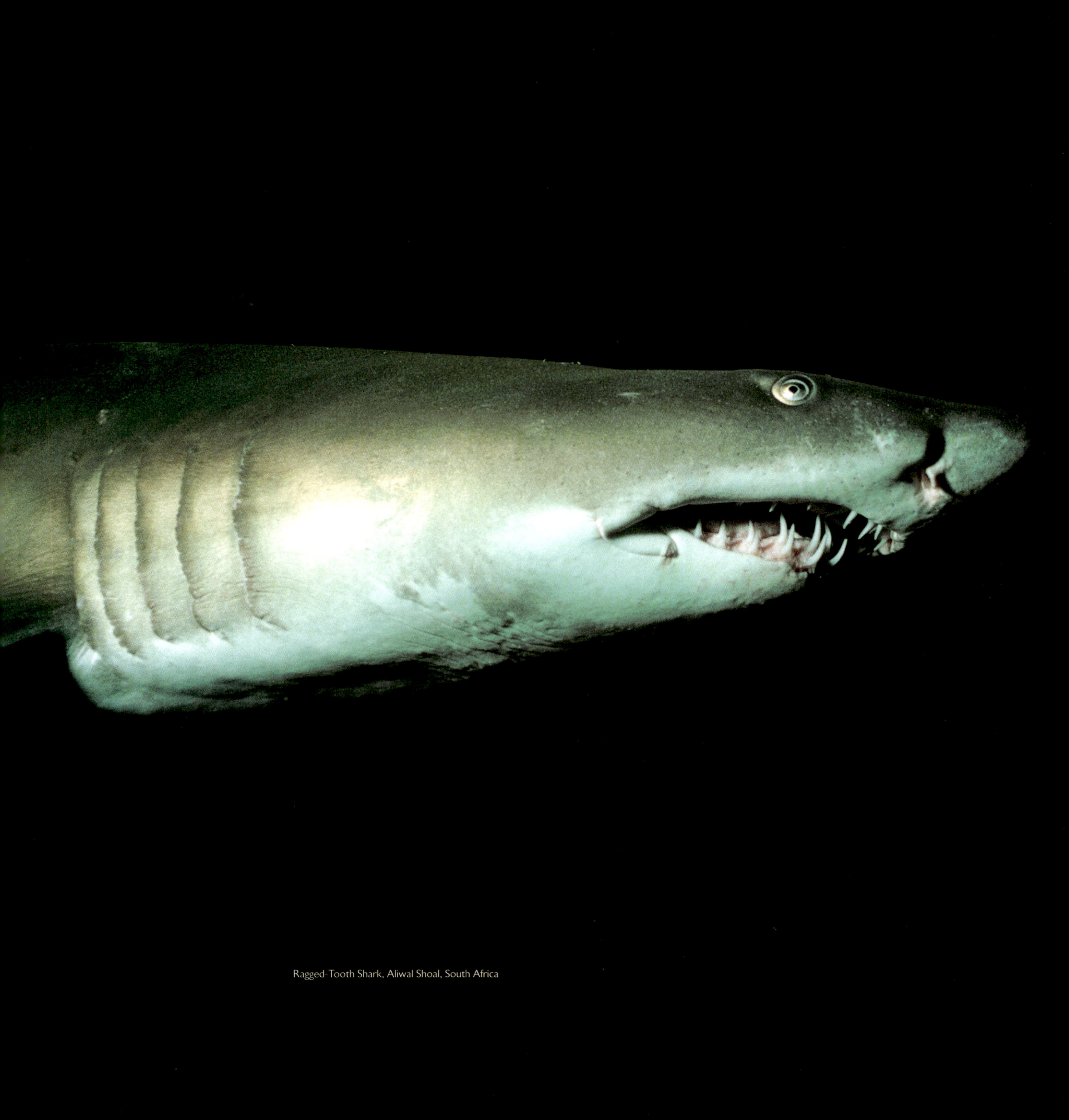

Ragged-Tooth Shark, Aliwal Shoal, South Africa

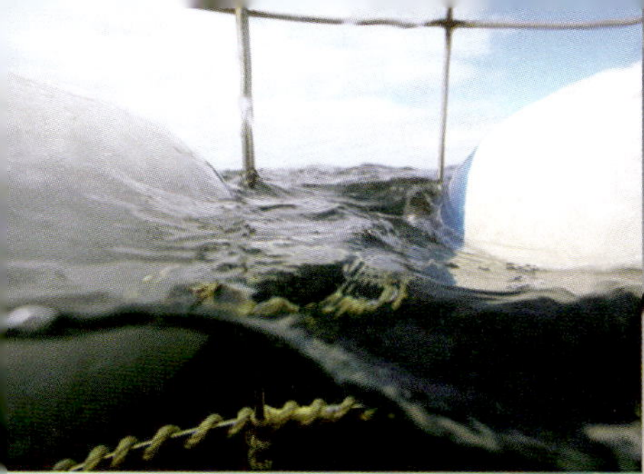
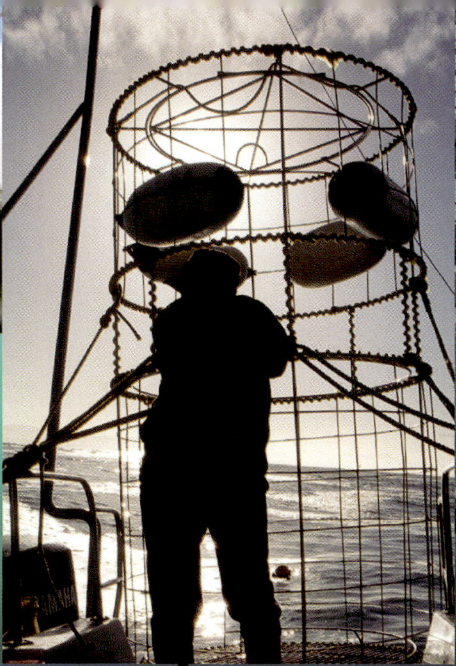
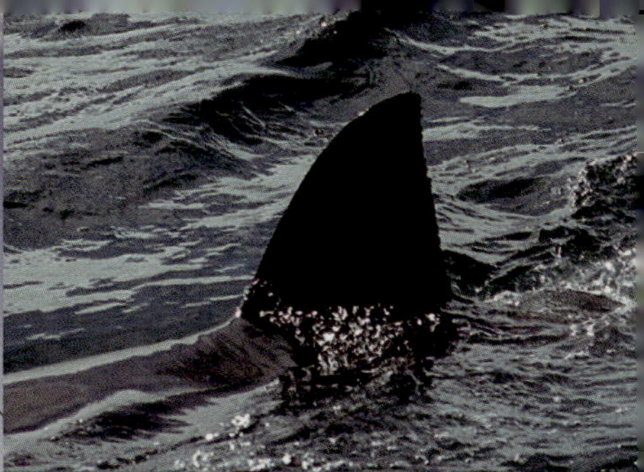
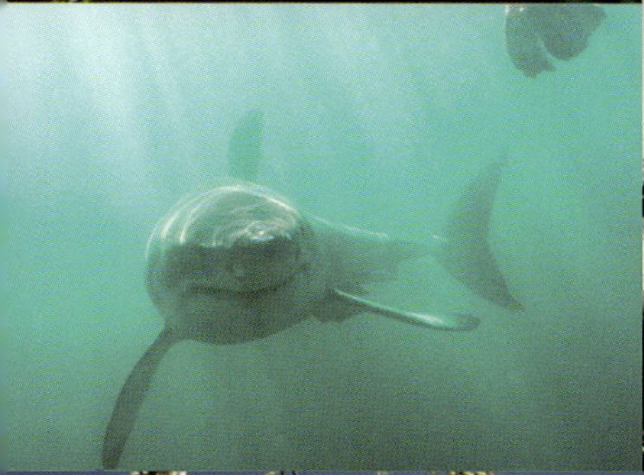
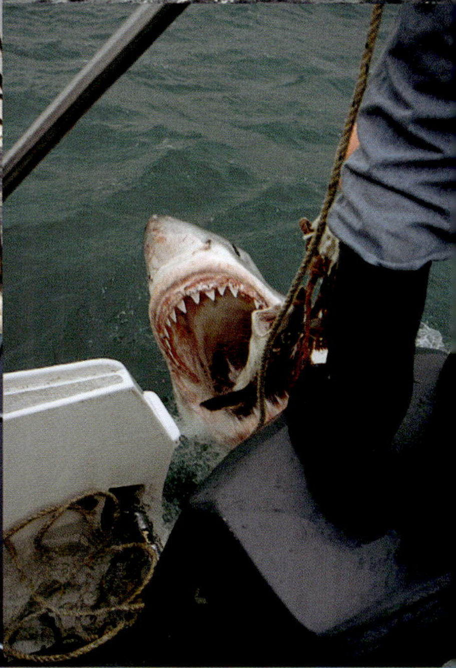
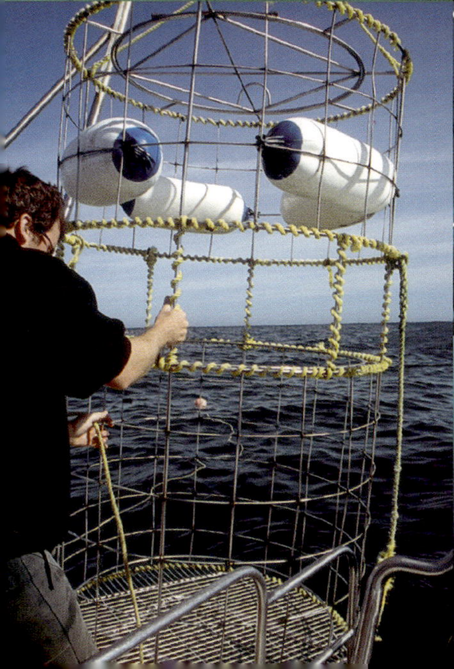
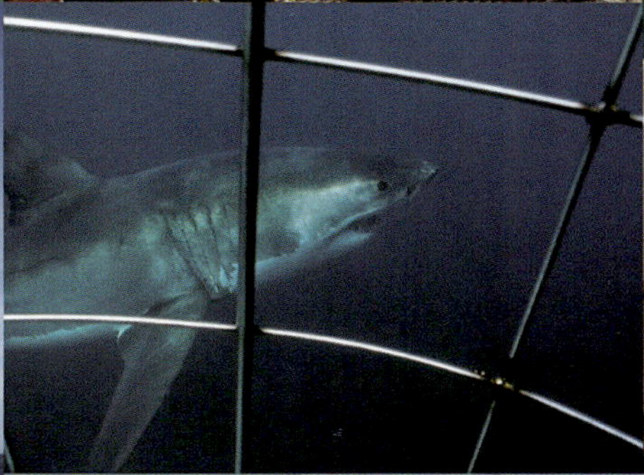
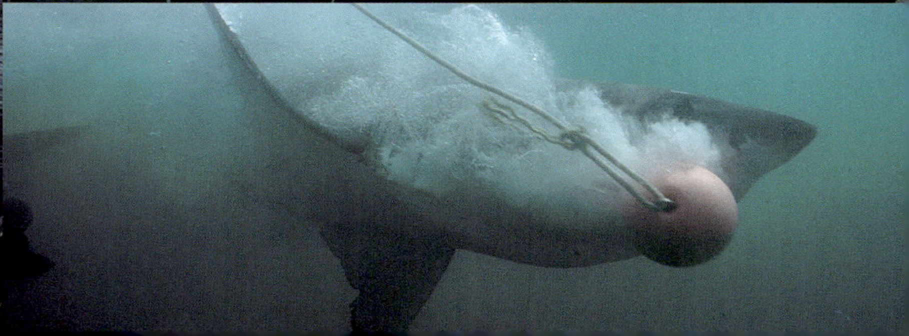

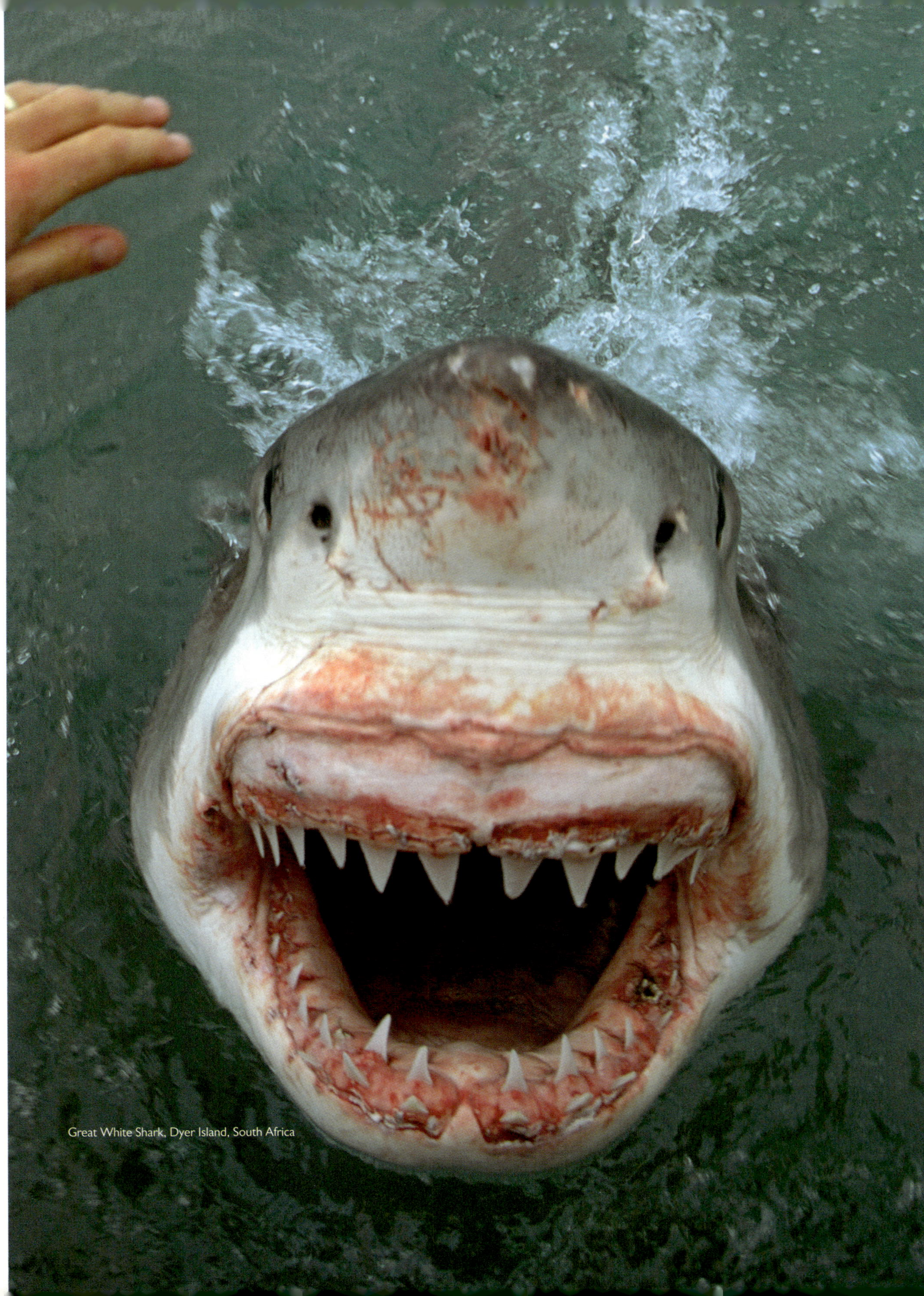

Great White Shark, Dyer Island, South Africa

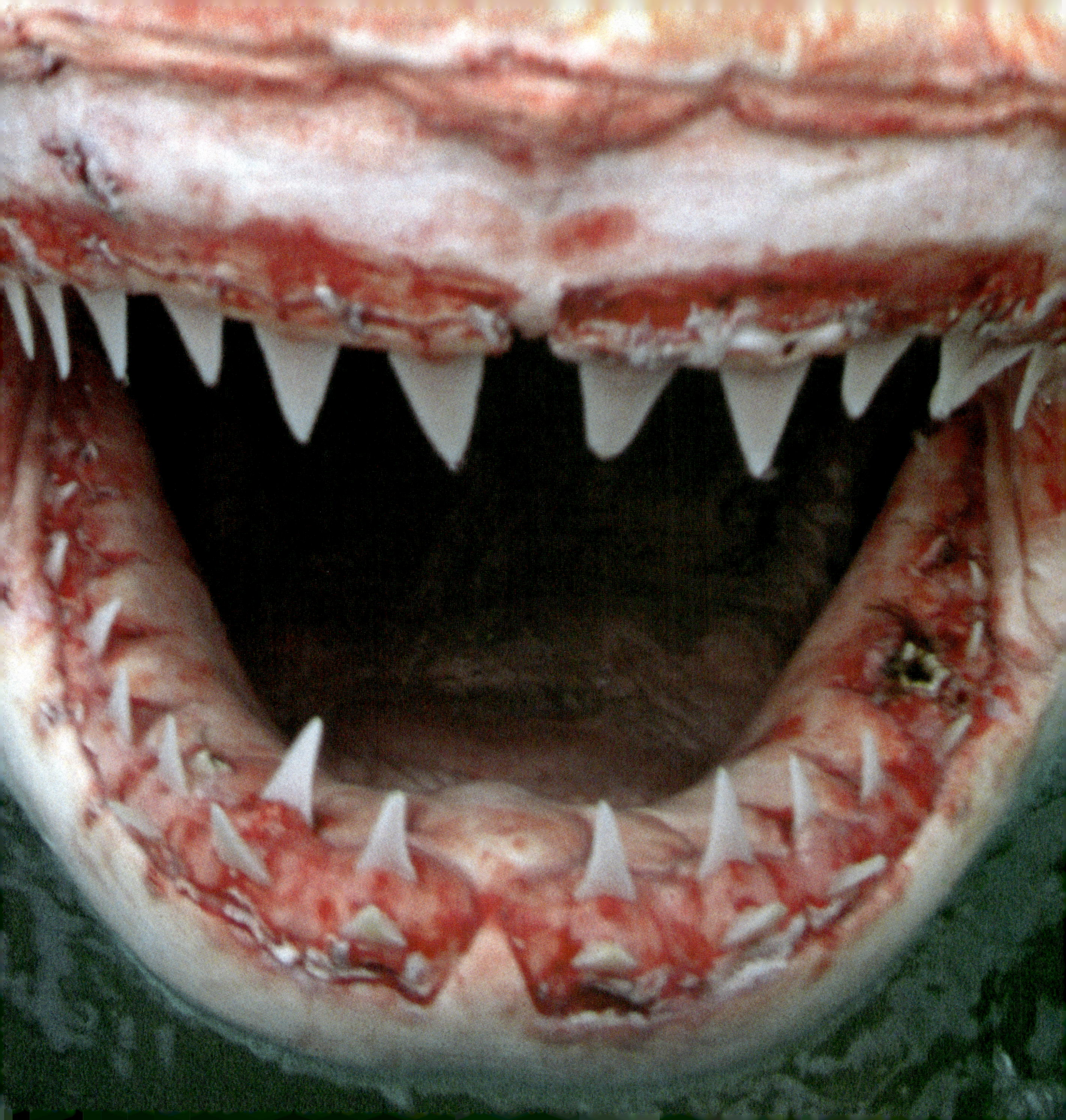

The water was cold, a spill from the chill of the Southern Ocean that traversed the bottom of the world, and my core body heat was dropping; it could no longer warm the seepage penetrating my neoprene wetsuit. I shivered, and my teeth chattered against the rubber mouthpiece of my regulator…

I thought of the animal I was there to see: the great white shark, largest of all the carnivorous fish in the sea. Rarely had it been seen underwater; rarer still were motion pictures of great whites in the wild.

And I thought of why I was bobbing alone in a flimsy cage in the frigid sea: I had written a novel about a shark, and had called it 'Jaws', and when it had unexpectedly become a popular success, a television producer had challenged me to go diving with the monster of my imagination. How could I say no?

Now, though, I wondered how could I have said yes.

Peter Benchley, *Shark Trouble*, 2002

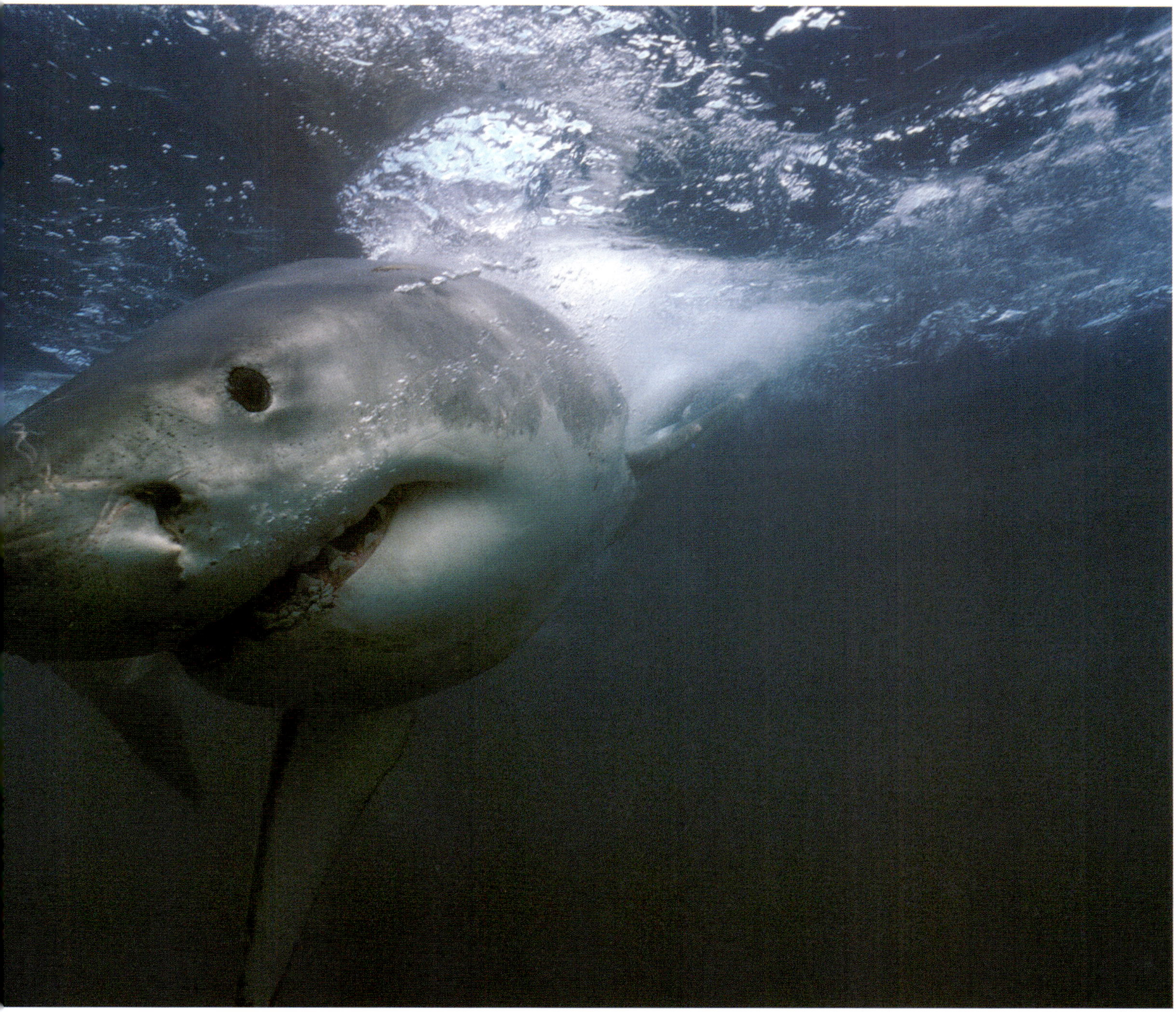

Great White Shark, Dyer Island, South Africa

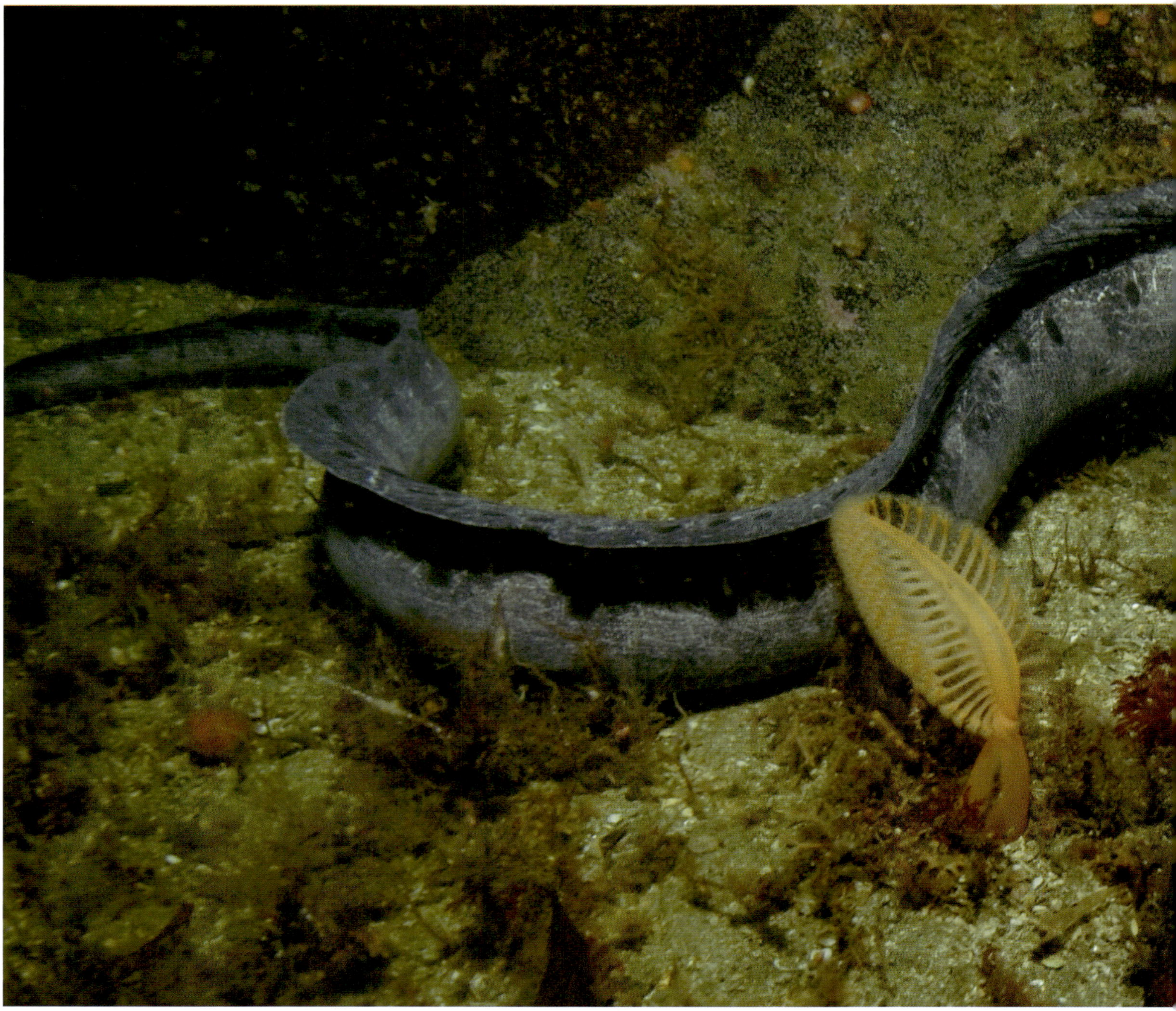

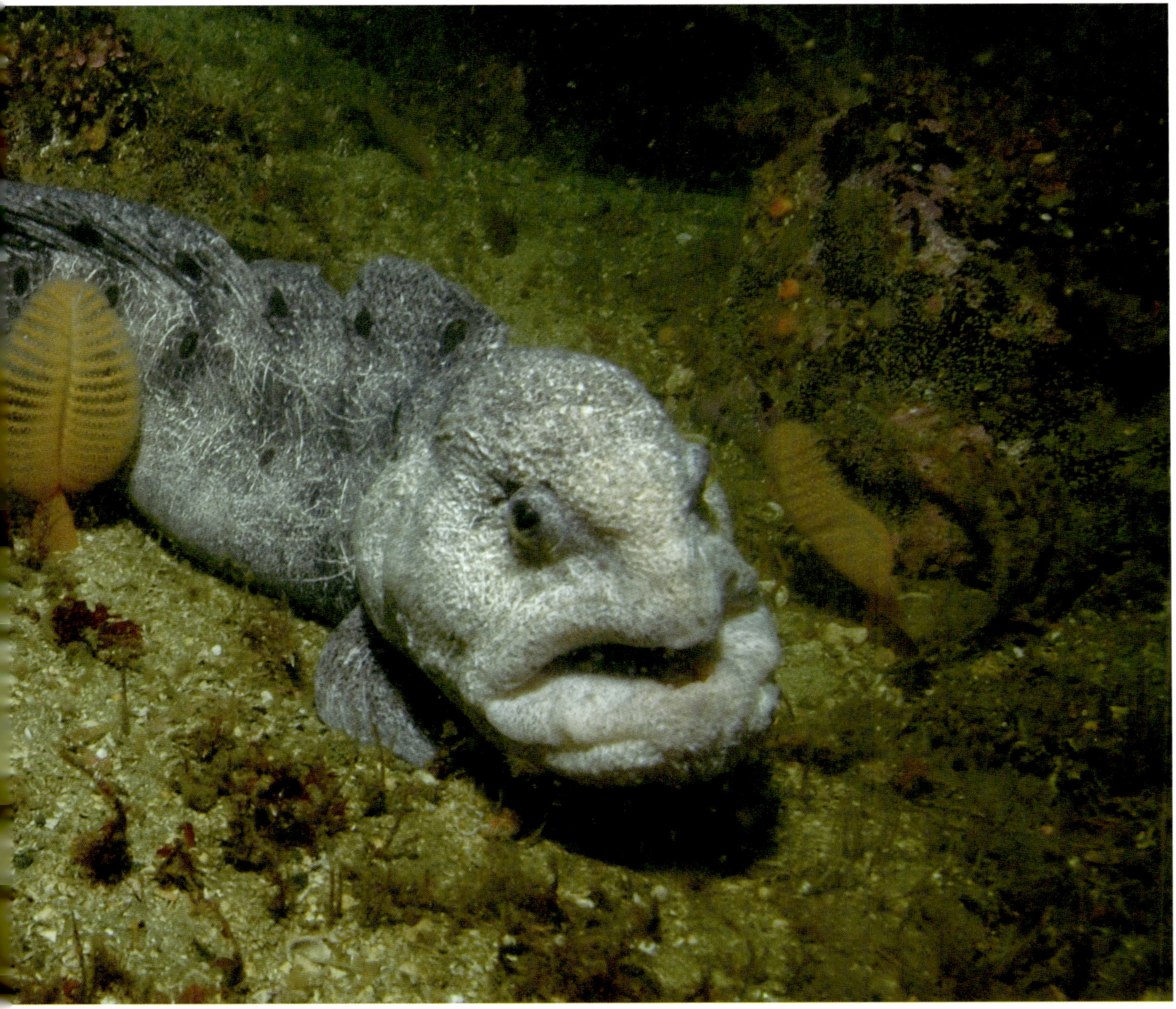

Wolf Eel, Dillon Rock, Vancouver

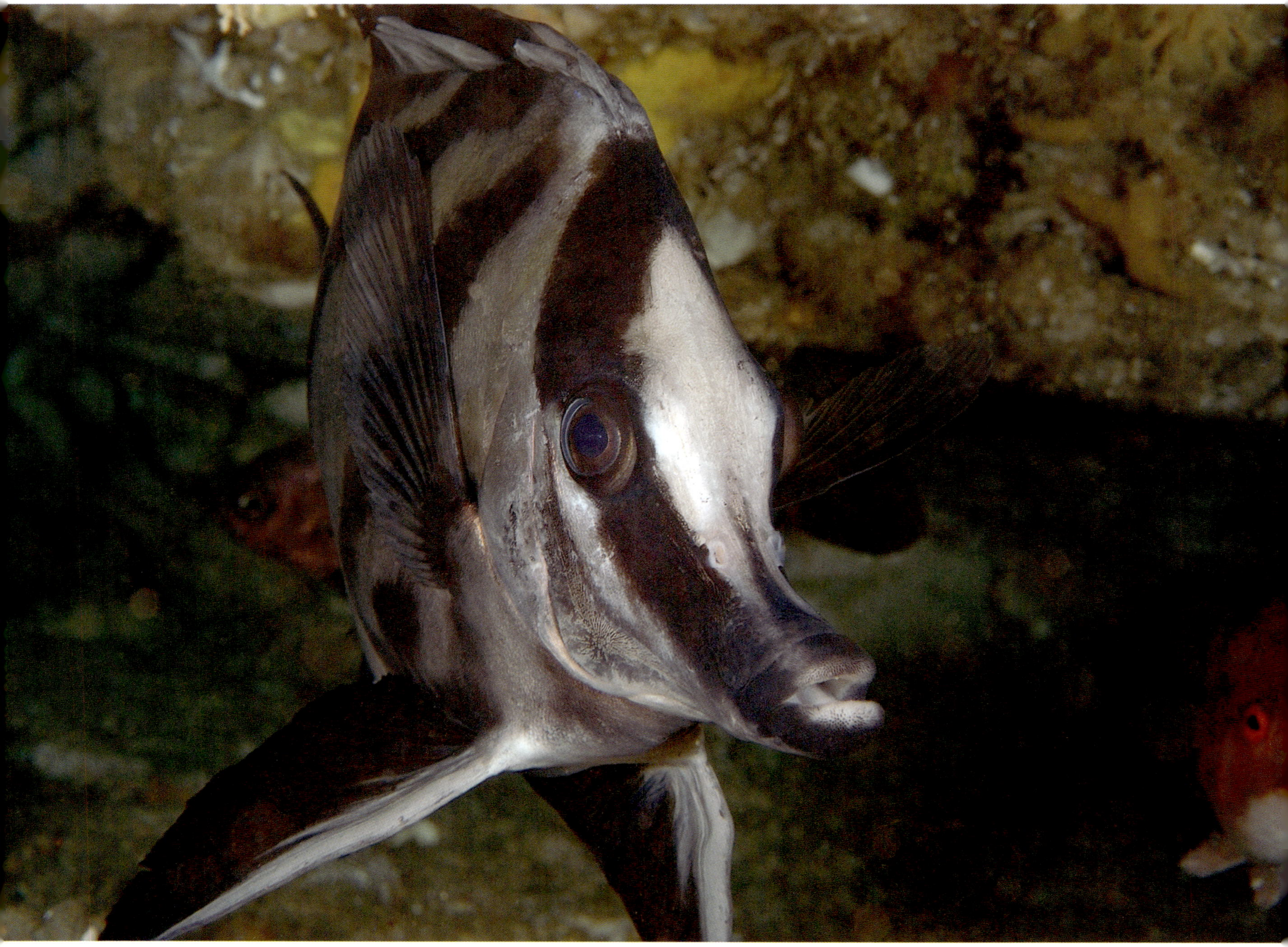

Bannerfish, Waterfall Bay, Tasmania

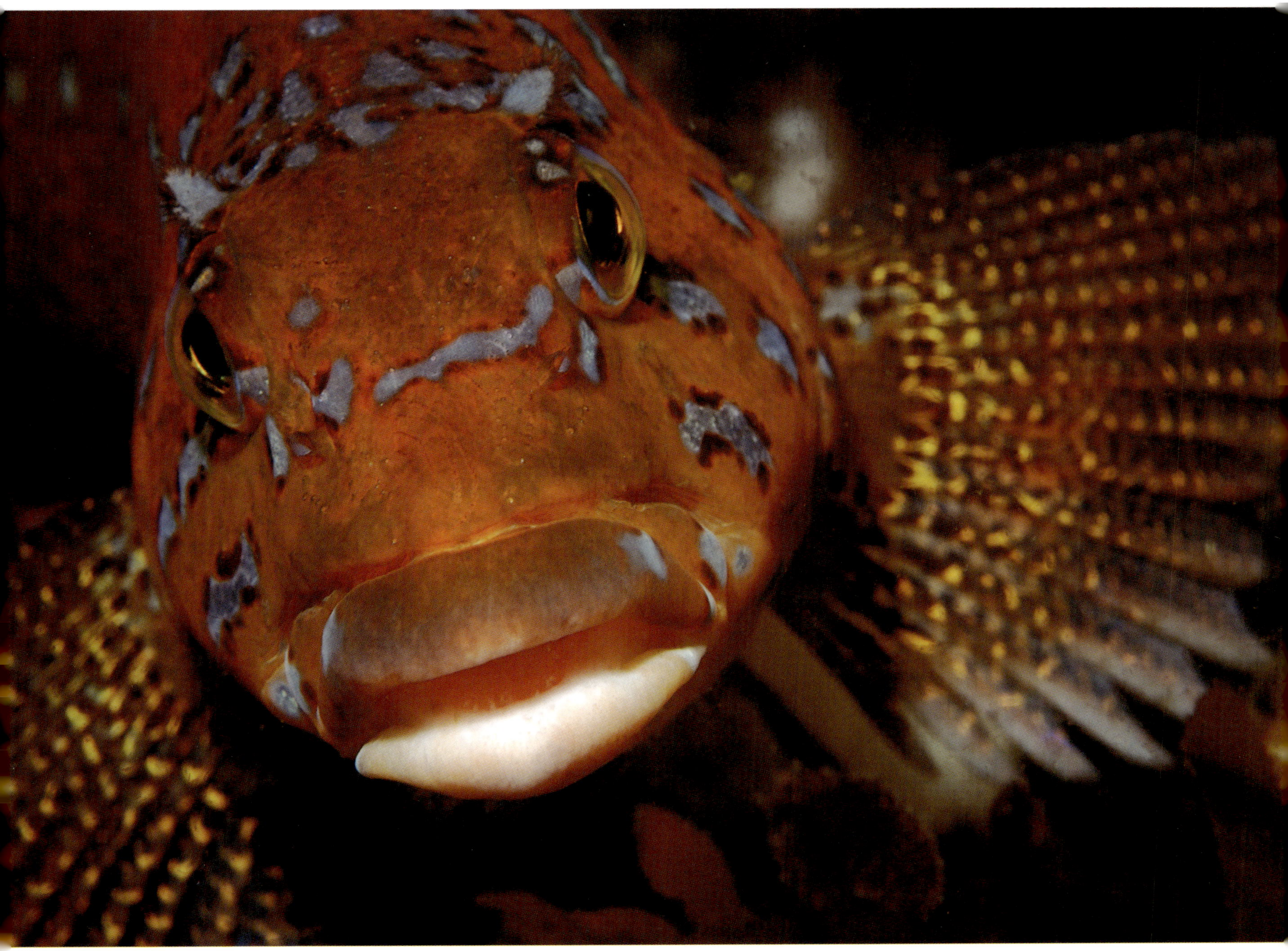

Kelp Greenling, Browning Pass, Vancouver

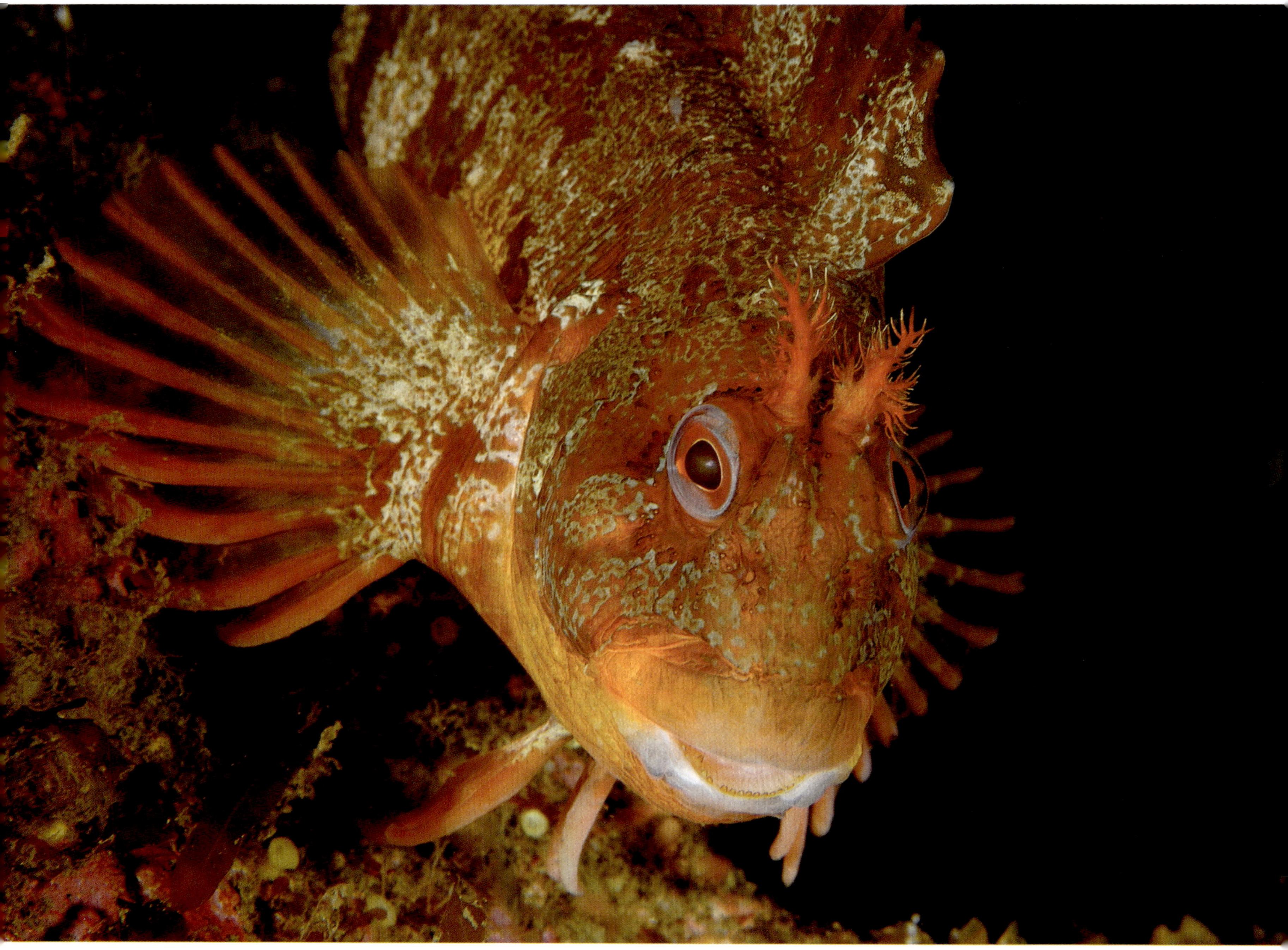

Tompot Blenny, Aran Islands

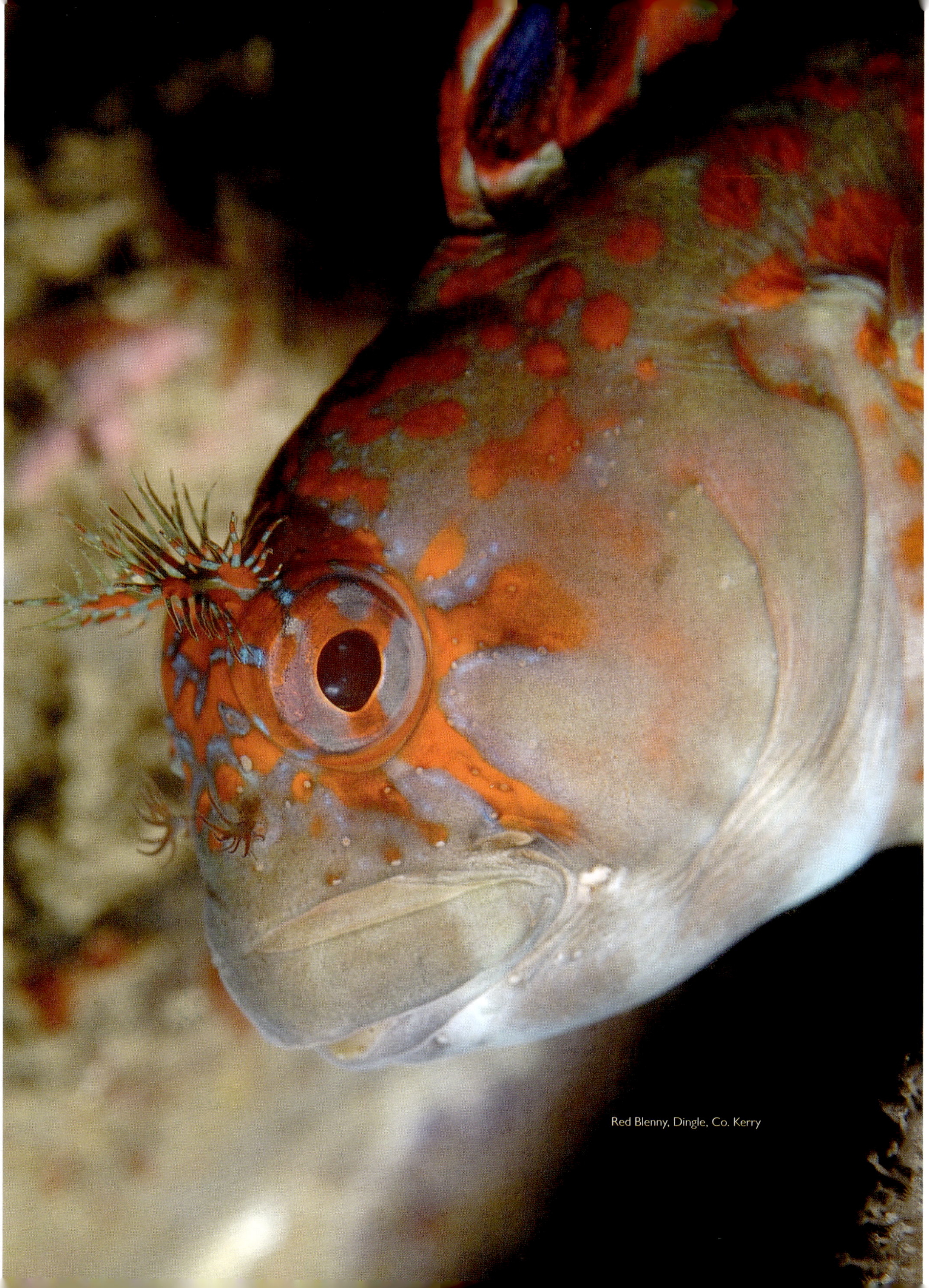

Red Blenny, Dingle, Co. Kerry

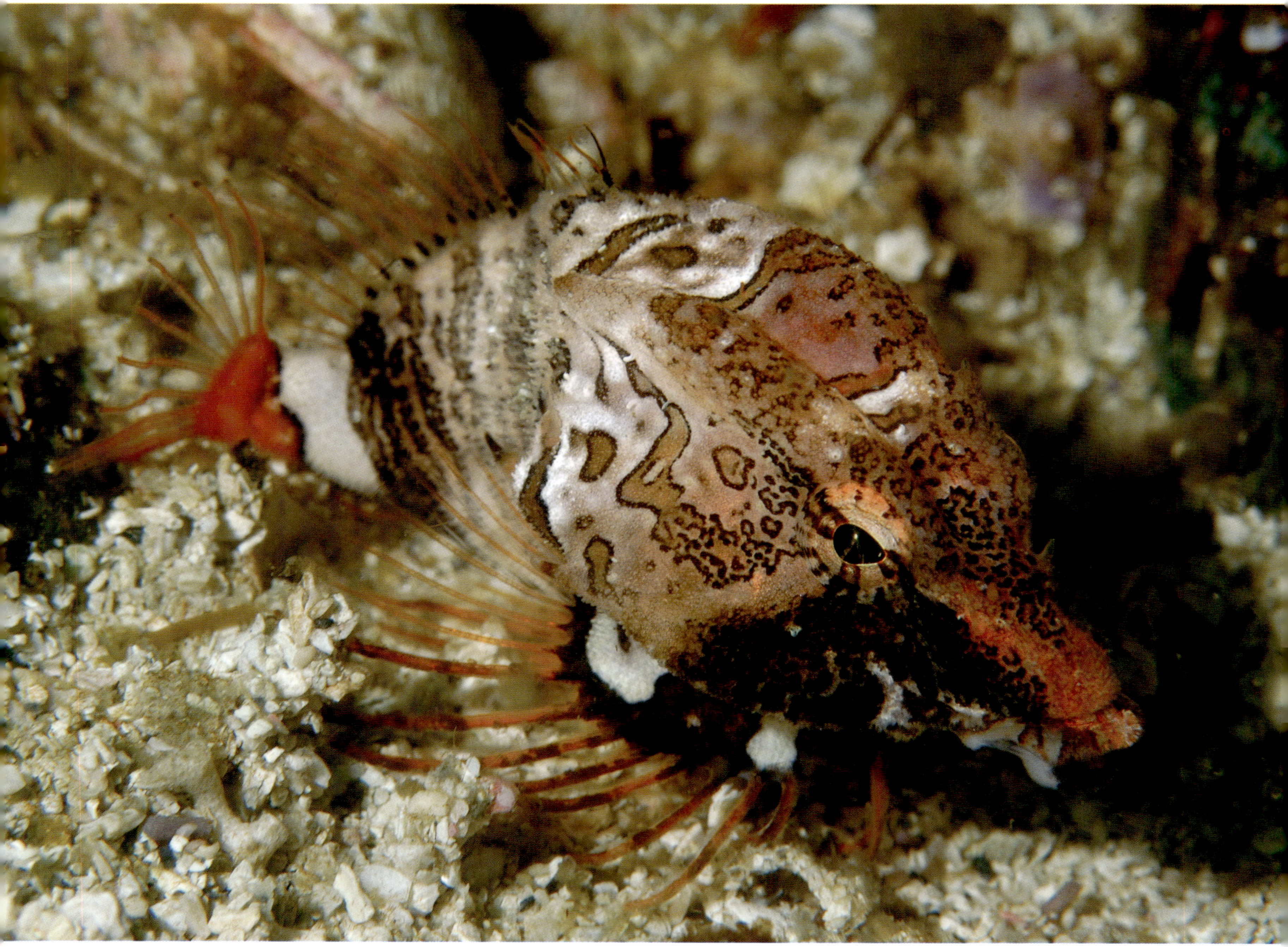

Grunt Sculpin, Hayden Pass, Vancouver

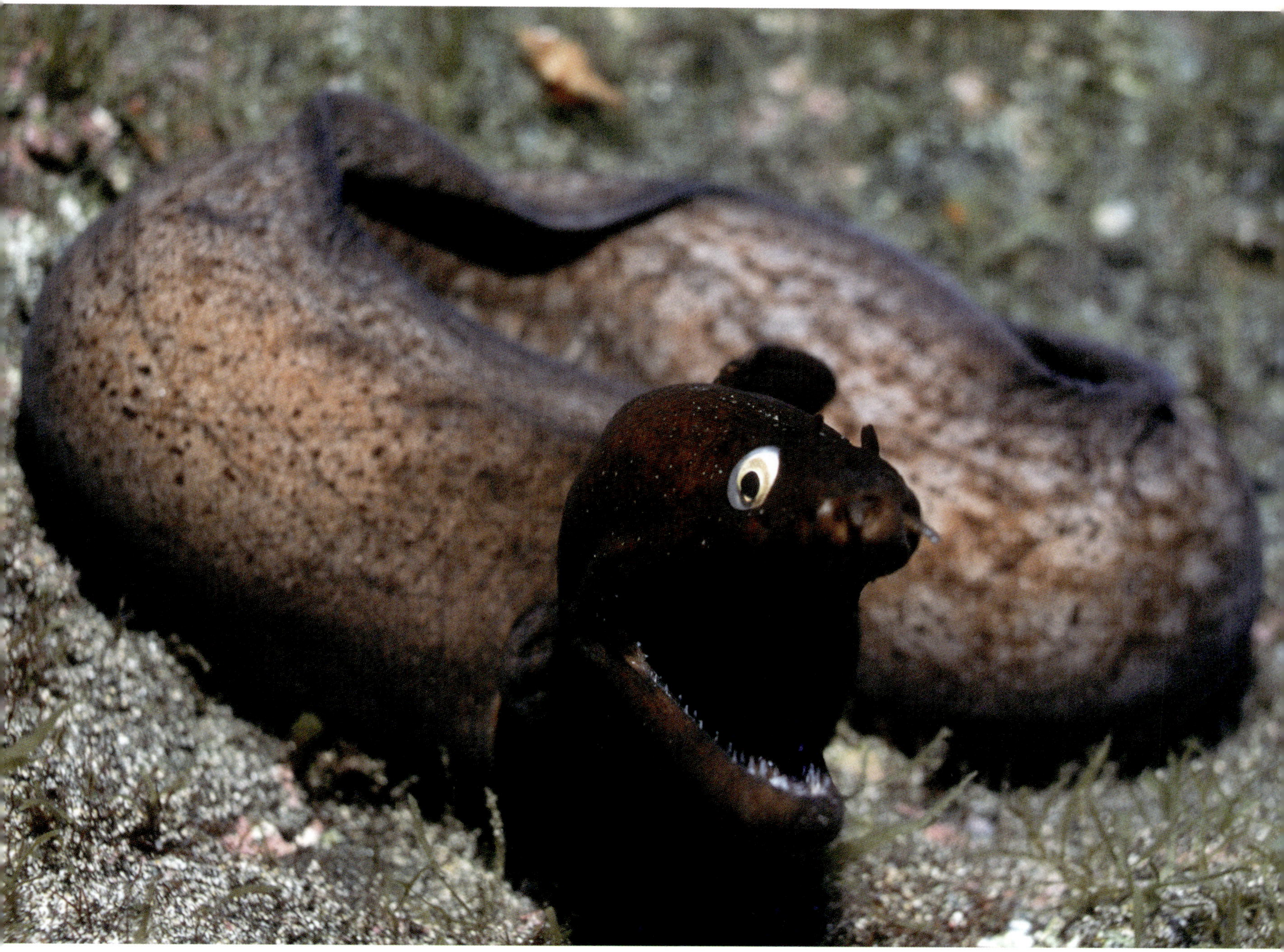

Juvenile Moray Eel, El Hierro, Canary Islands

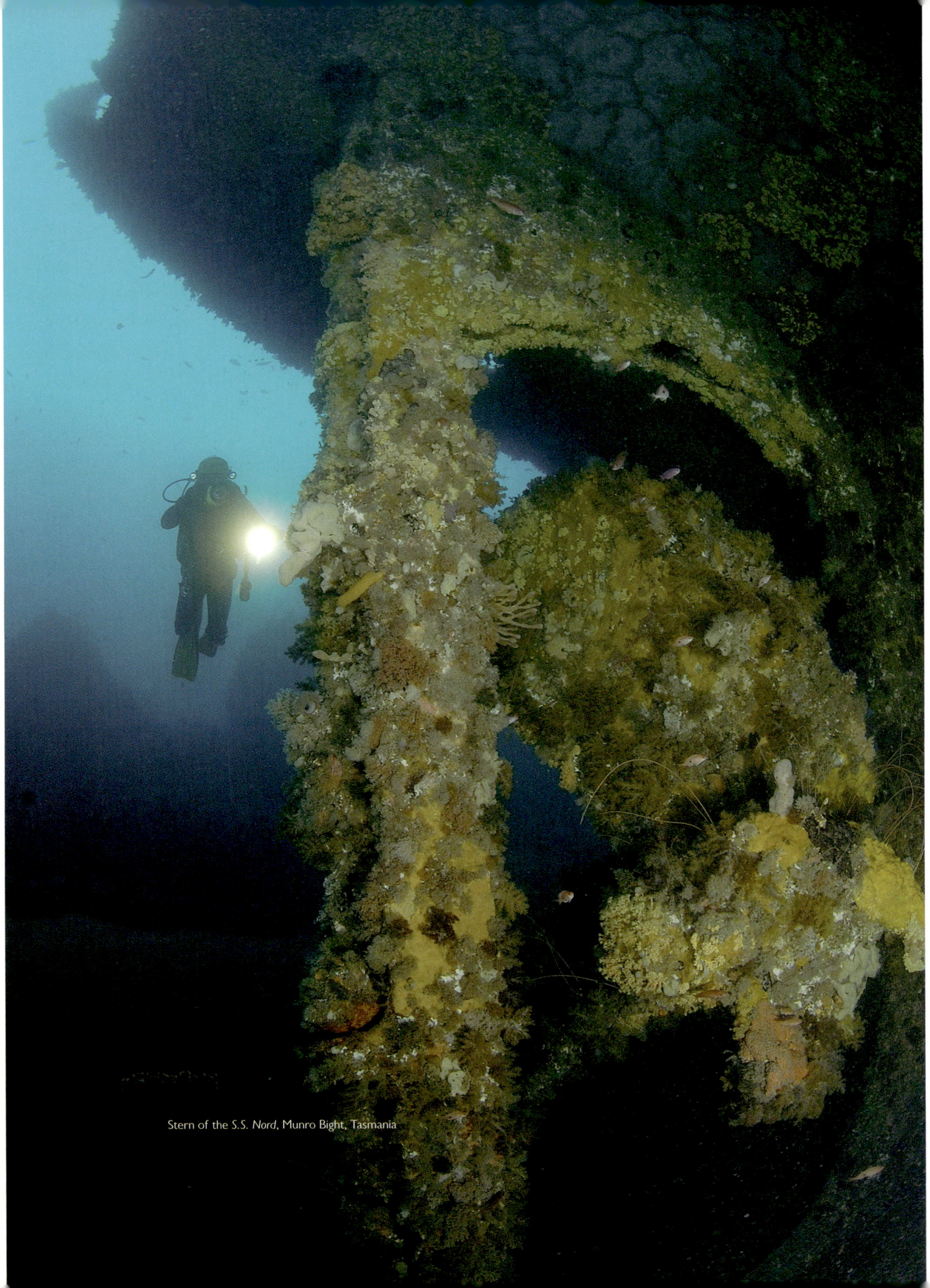

Stern of the S.S. *Nord*, Munro Bight, Tasmania

WRECKED AT SEA

Shipwrecks have a special attraction to us. Perhaps it is the sense of mystery born out of childhood stories of sunken treasure. Certainly, there is a fascination with the loss of a once proud ship – even if its history is well known. Some elements of a wreck are immediately recognisable but mostly, as they are consumed by the sea, only a jumble remains. Swimming among wreckage is often a puzzle to both the eye and the imagination. On a deep wreck dive off the Devon coast, I recall hovering over the ship's railing, wondering about who might have last stood there, and being suddenly transported back to one of my earliest childhood memories. Crossing the Irish Sea, returning home from England, I had somehow momentarily escaped my parents' clutches. I had been innocently playing with some plastic cutlery and got a notion to see them fly into the water below. Slipping out the doorway, I managed to climb a couple of rungs of the railing and watched, transfixed as my white spoons disappeared into the frothy wake below. Suddenly, I was grasped by an anxious father and the game was up. My mind returns to the dive at hand and I check on depth, time and air supply – the fundamental routines of diving. It is time to start the swim back to the shot-line and begin the slow ascent back to the light.

Soon after sinking, a wreck becomes part of the underwater landscape. However unwelcome, these intruding hulks will become one with the sea. Within days, animals will adopt it as a new home. As the seasons pass, the crumbling structures become an artificial reef with all of the life of its rocky neighbours. Shelter for schools of fish, conger eels, lobster and crawfish. So, to a diver, it appears to have all of the life and colour of a rocky reef, but now with an added dimension. There is a separate sense that this was once part of our world, made by us, but now taken by the sea.

On deeper dives, air is not a perfect breathing gas for diving. The air we breathe on land serves us well, comprising of one-fifth oxygen and almost four-fifths nitrogen. The nitrogen fraction of air has no role in our earthly lives but as pressure increases rapidly at depth, nitrogen starts to dissolve into our bodily tissues where its effects are profound. The first effect was described by Jacques Cousteau as "l'ivresse des grandes profondeurs" (rapture, or intoxication, of great depths), now known as Nitrogen Narcosis. It is a type of drunkenness that appears at depths greater than thirty metres. The affected diver loses co-

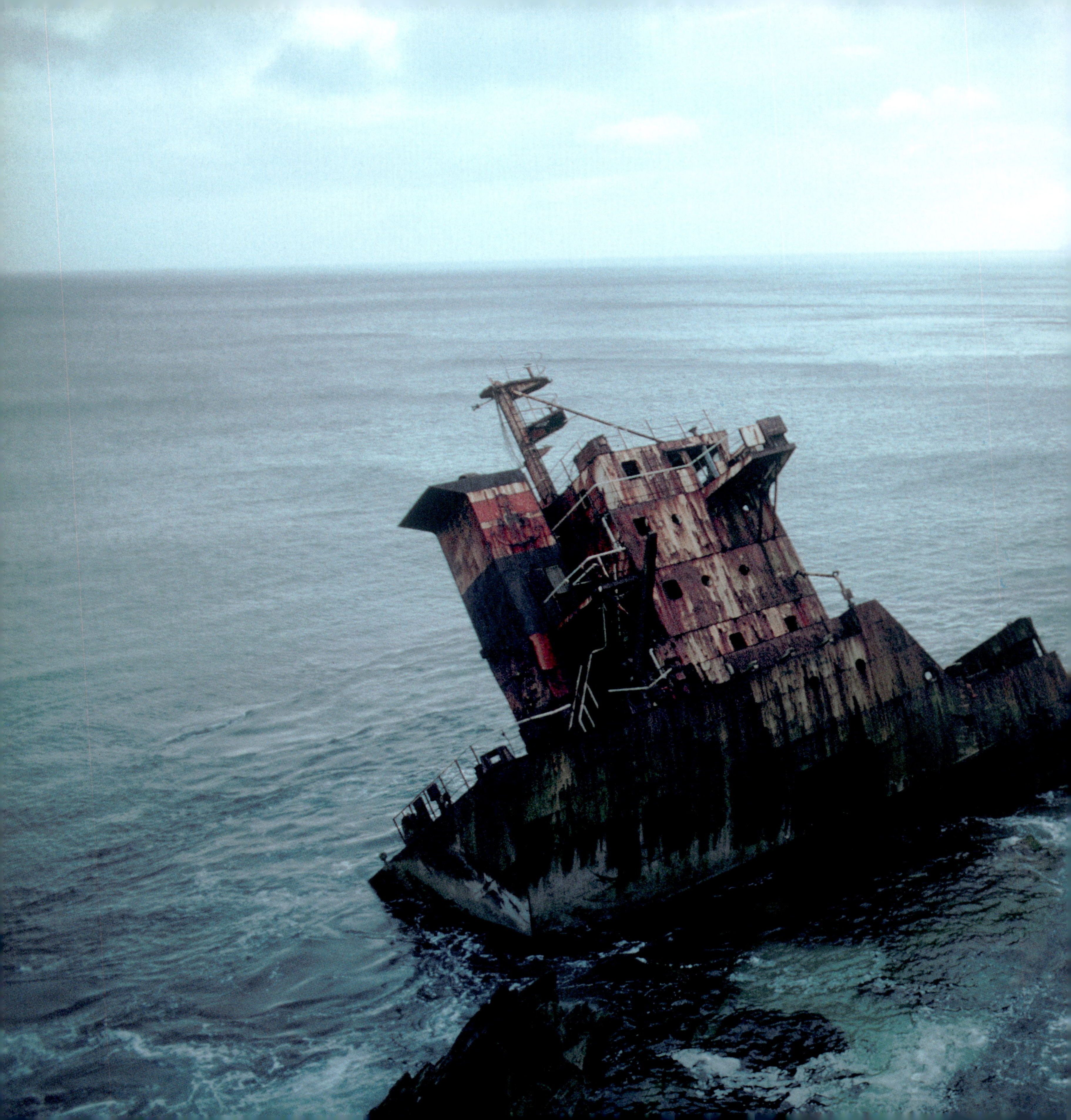

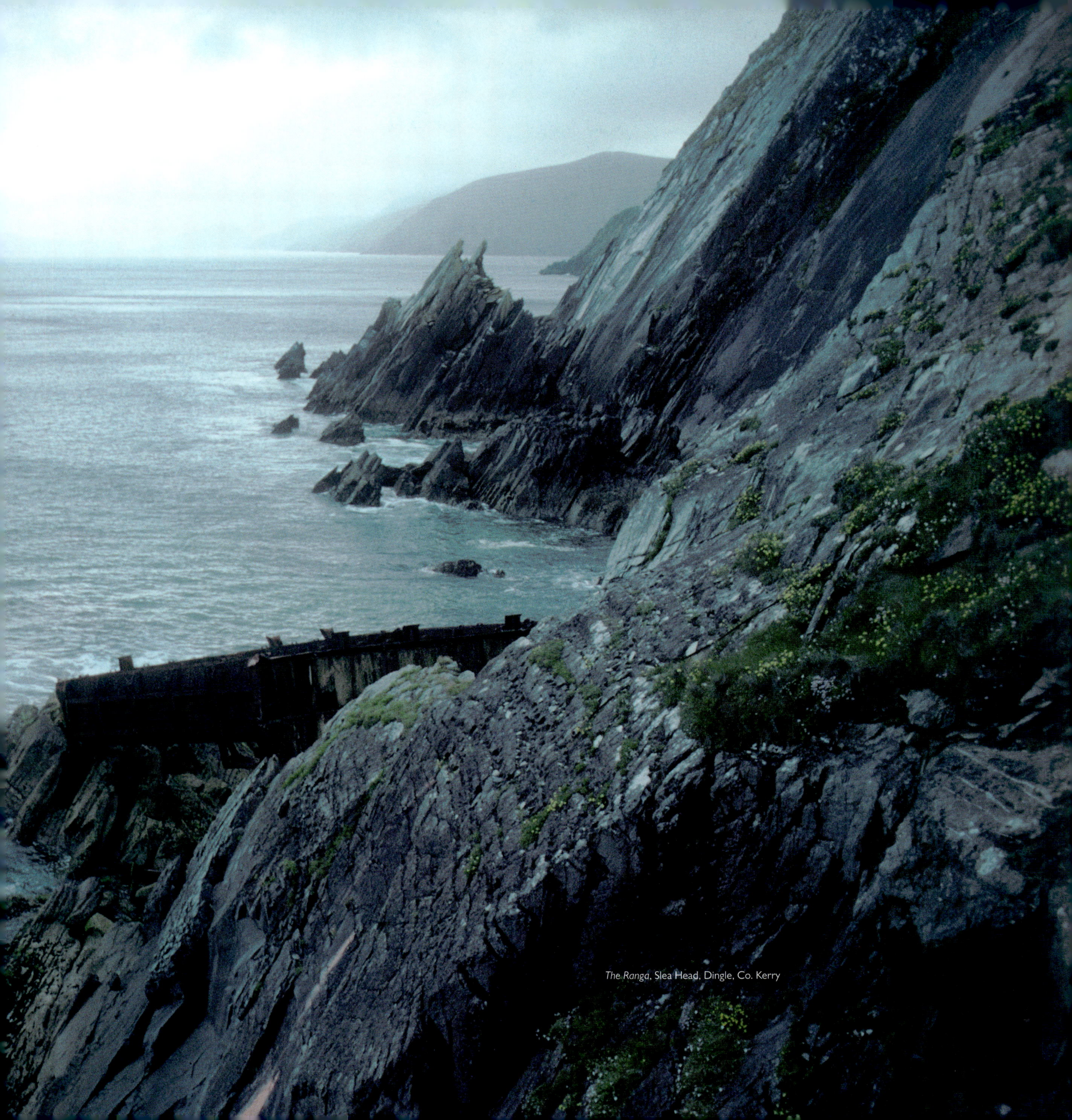

The Ranga, Slea Head, Dingle, Co. Kerry

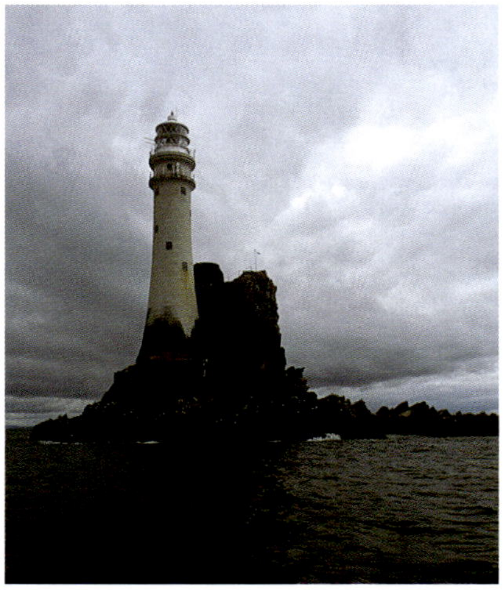

Carraig Aonair, the Fastnet Rock

ordination, can feel dizzy or light-headed or may even offer his air supply to passing fish. Some will feel euphoric while others, particularly on deep, dark dives will become increasingly fearful and paranoid. Whatever the subtle differences, the fundamental effect of 'the narc's' is to impair the diver's judgement.

The second effect of breathing nitrogen at depth is the dreaded 'bends' or Decompression Illness. This painful and potentially fatal affliction is a result of breathing compressed air at increased pressure and rapidly returning to the surface. It was first seen in caisson workers in the late nineteenth century, during the construction of tunnels. While excavating under rivers, work areas were supported by pumping in compressed air and as workers emerged from their shifts, many became suddenly ill. The most common symptom was severe pain in the joints which worsened on straightening the affected limb but was relieved by bending it – hence 'the bends'. We now know that decompression illness is a result of nitrogen bubbles forming in the tissues and that dissolved nitrogen built up during a dive has to dissipate gradually during ascent. The limits of depth and time, as well as a slow swim to the surface are fundamental to safely diving deeper wrecks.

Late in World War II, the young crew of a German submarine climbed the rocks of Galley head in West Cork, grateful to have made it ashore. Their U-boat, U-260, had struck a mine near the Fastnet Rock and was taking on water. As the banks of batteries were swamped, deadly chlorine gas was produced and they were ordered to abandon and scuttle their vessel. It now lies just off Glandore, in forty five metres and is a necessarily short dive, undertaken with great care. On descent, as the light fades, it looms out of the darkness, large and bold. Landing on top of the sleek vessel, it is initially difficult to get your bearings but a short swim over the side shows that it is lying almost on its side. Various pipes and valves are visible on the deck, once covered with polished timber which has long since rotted away. Suddenly, the conning tower, centre of all life aboard a submarine, appears in the torch beams and is mightily impressive. The hatch remains open from where the last sailor escaped, the attack and aircraft periscopes are intact with glass and gears visible. It is easy to imagine an eye at the other end, searching the horizon. Fifteen minutes pass as if only two or three and the ascent to the surface must now begin, without delay.

Such visits back in time are special. The evidence of a turbulent time in history is even more tangible on the wreck of the 'Aud' outside Cork harbour, again on Ireland's south coast. Here, the seabed is covered with thousands of bullets destined for the 1916 Rising against British rule in Ireland. In the tradition of acquiring help from Britain's enemies, Germany was pressed and agreed to support an armed insurrection by supplying 20,000 rifles and a million rounds of ammunition. These were loaded onto a captured British steamer, the 'SS Castro', which had been renamed the 'Libau' and was now carefully camouflaged to look like an existing Norwegian ship, the 'Aud'. Despite a severe storm off Rockall, the German naval crew reached Tralee Bay ahead of schedule but unfortunately did not receive the arranged signals to attempt landing the arms. The following day the suspicions of the authorities had been raised and the 'Aud' was soon arrested by a British warship. Rather than let the ship and its cargo fall into enemy hands, the captain ordered the ship scuttled while being escorted to Cobh, then known as Queenstown. The wreck now lies in thirty six metres of water, just outside Cork harbour and is a fascinating journey through the pages of history.

Little remains of old wrecks. Exposed wooden structures rot quickly if not buried in sand or mud. Those near the coastline face further destruction during winter storms when waves pound the wreckage into the seabed. After hundreds of years, the remaining pieces of recognisable wreckage are heavily encrusted with marine growth, disguising them even more. A lone anchor in a shallow area of seaweed is sometimes all that remains of a once magnificent tall-ship. We can but wonder how these ships once looked, proud on their launch day as they faced their many adventures.

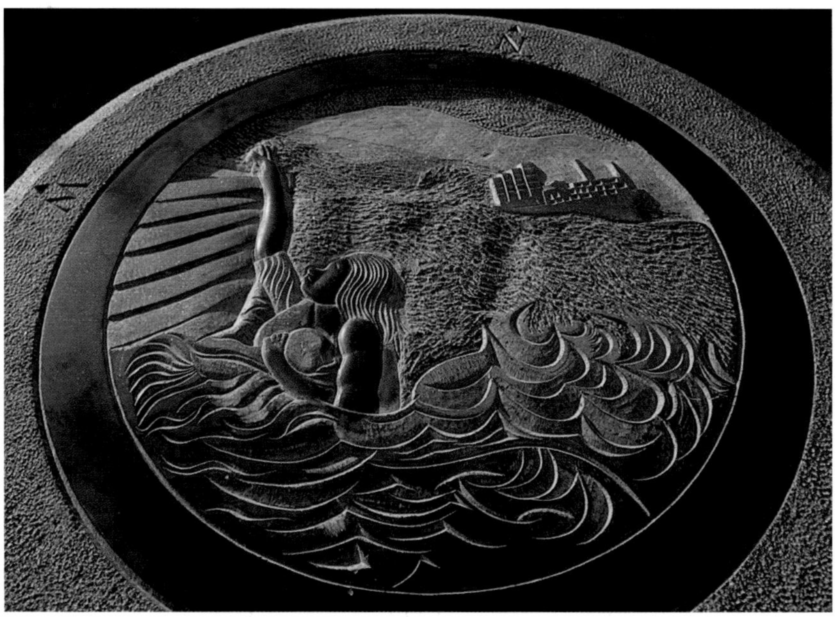

Lusitania Memorial, Old Head of Kinsale

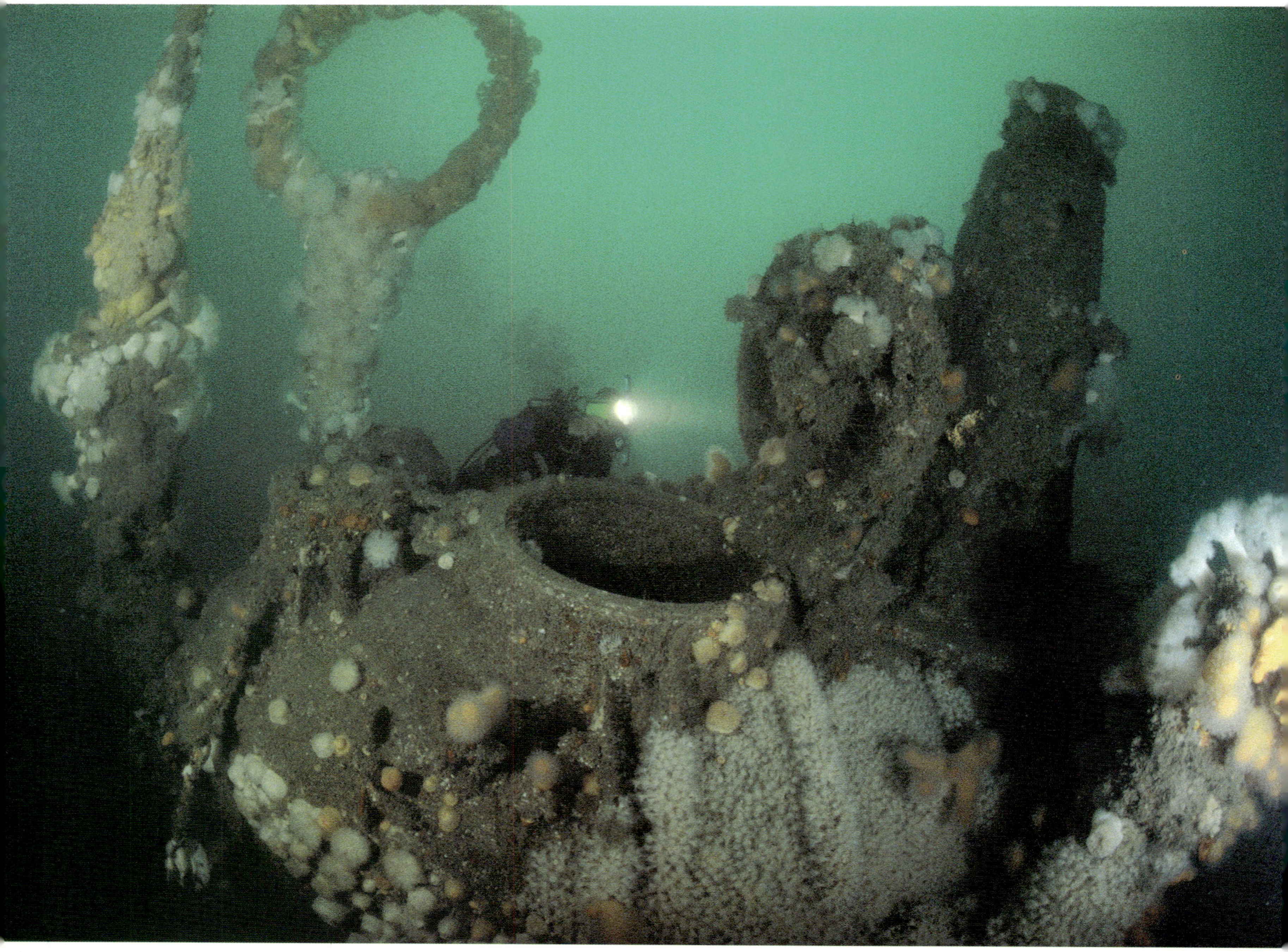

Conning Tower, *U-260*, Glandore, Co. Cork

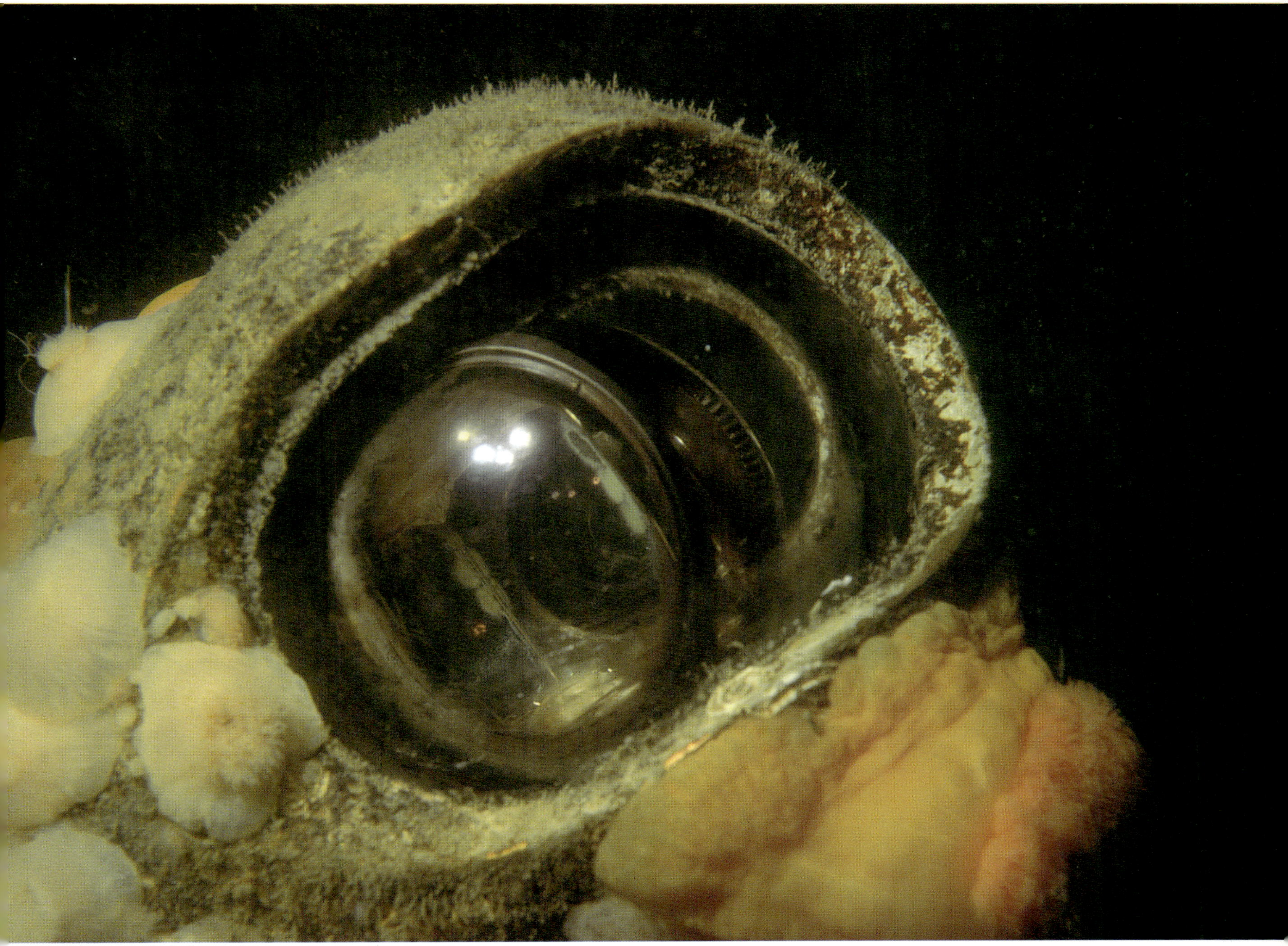

Periscope, *U-260*, Glandore, Co. Cork

On a single dive off Crete, Falco swam through perhaps four thousand years of history. At a depth of 125 feet he came upon a heap of spherical Byzantine amphorae and, along the slope fifteen feet deeper, another mound of cargo jars associated with an earlier Christian century. Around 150 feet down he found piles of bronze dishes. He did not disturb them, wishing to leave 'in situ' evidence for future archaeologists. At a depth of 170 feet Falco encountered a mass of round-bellied amphorae, identical with a type Sir Arthur Evans dug up at Knossos. The half-hour descent through liquid time – from 'Calypso' to the Bronze age – symbolised for us the lavish opportunity for archaeology in the Aegean.

Jacques Cousteau, *The Living Sea*, 1963

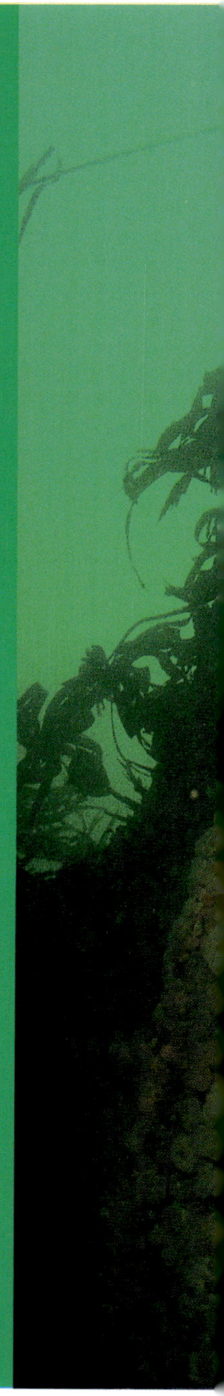

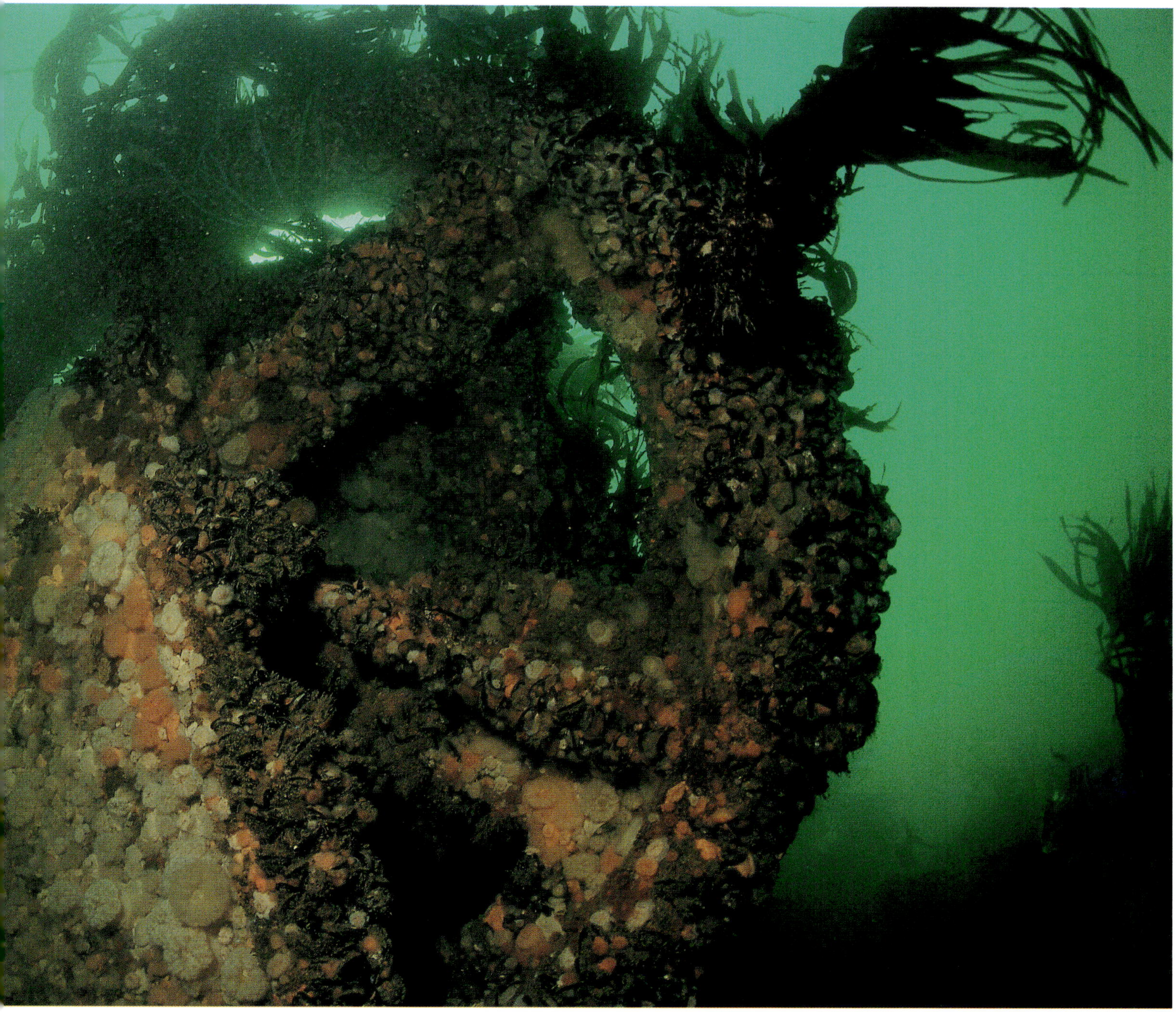

Winch, *Kowloon Bridge*, Stags Rocks, Cork

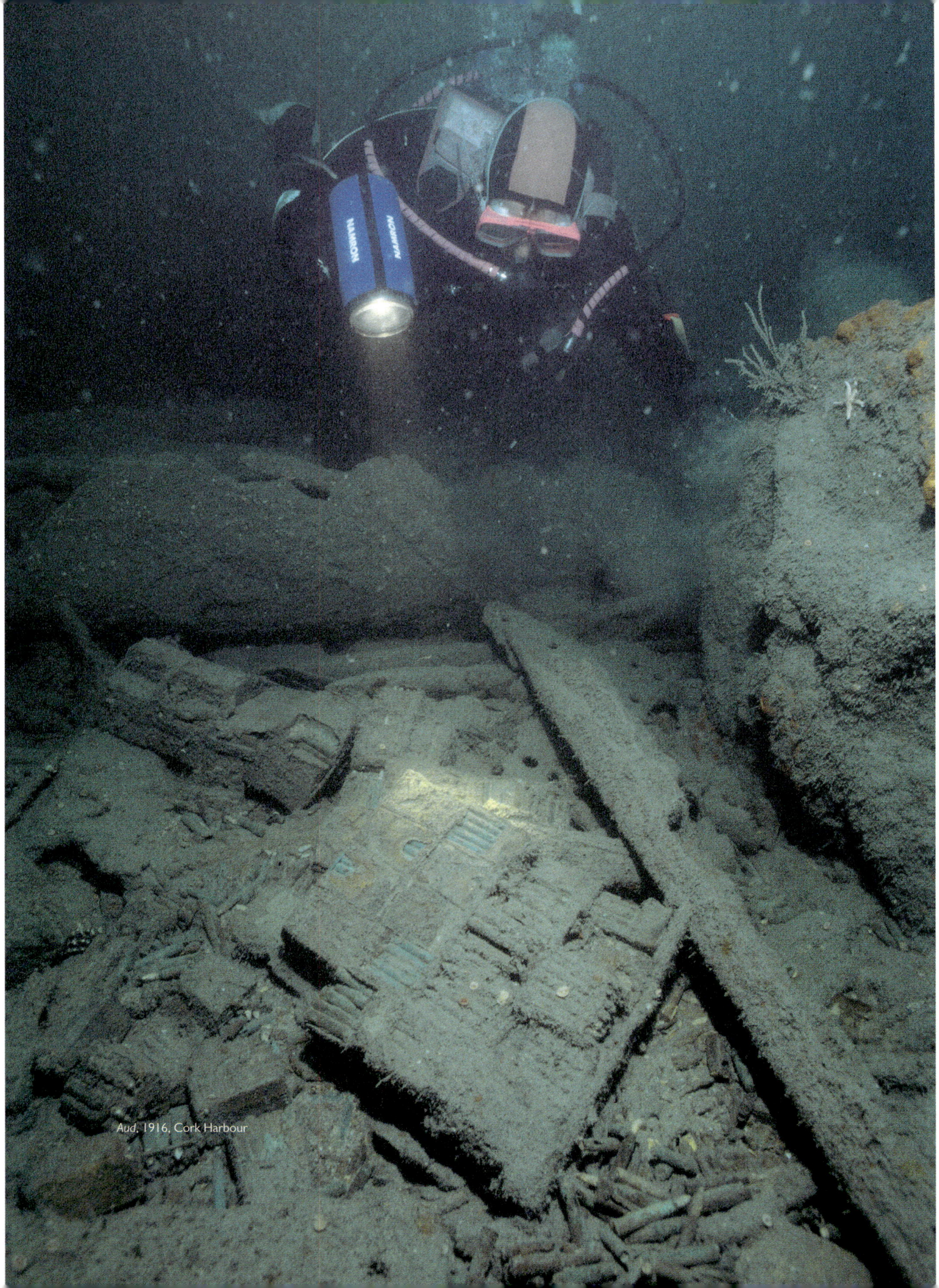

Aud, 1916, Cork Harbour

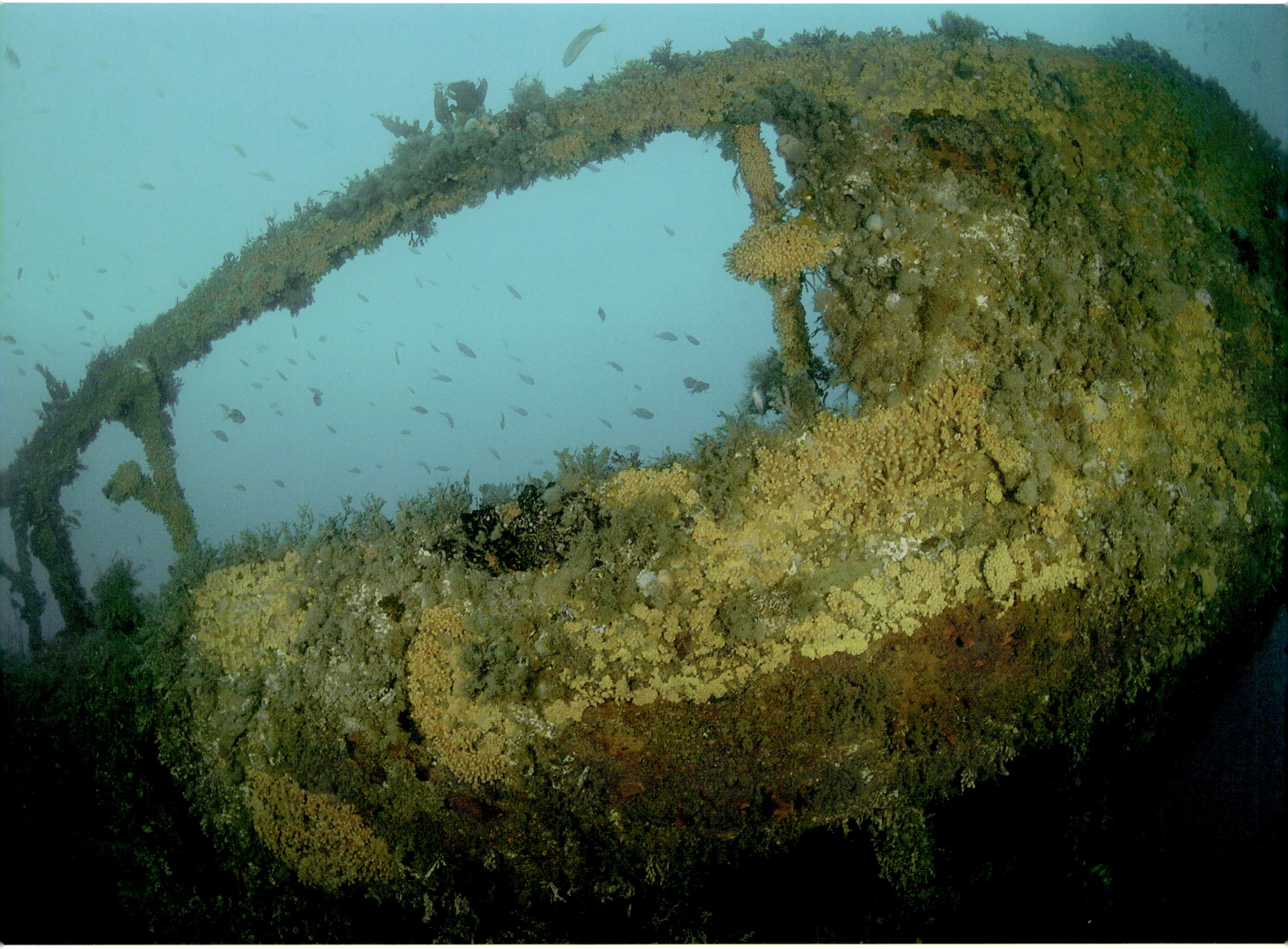

S.S. Nord, 1915, Munro Bight, Tasmania

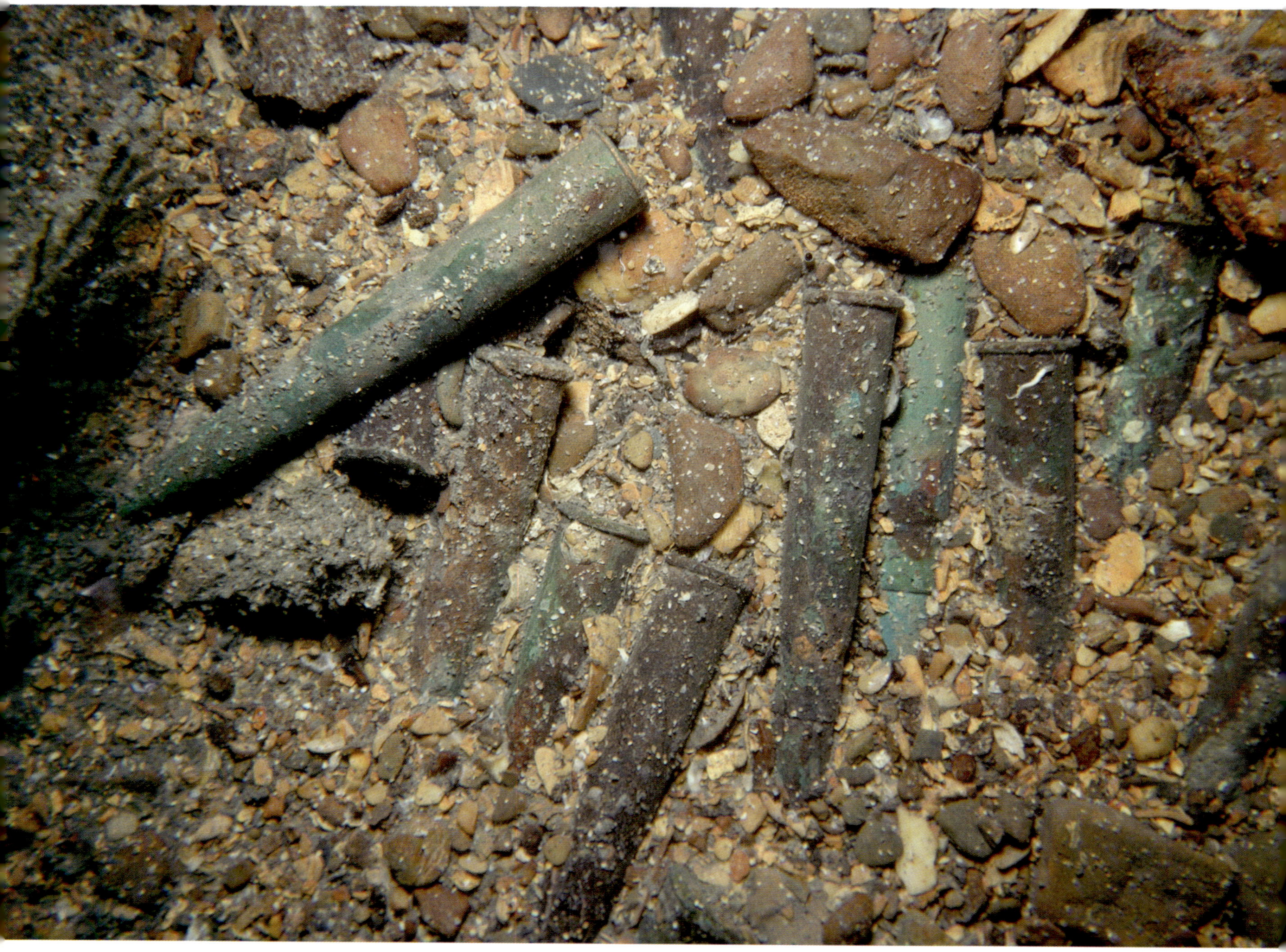

Bullets on the Wreck of *The Aud*, Cork

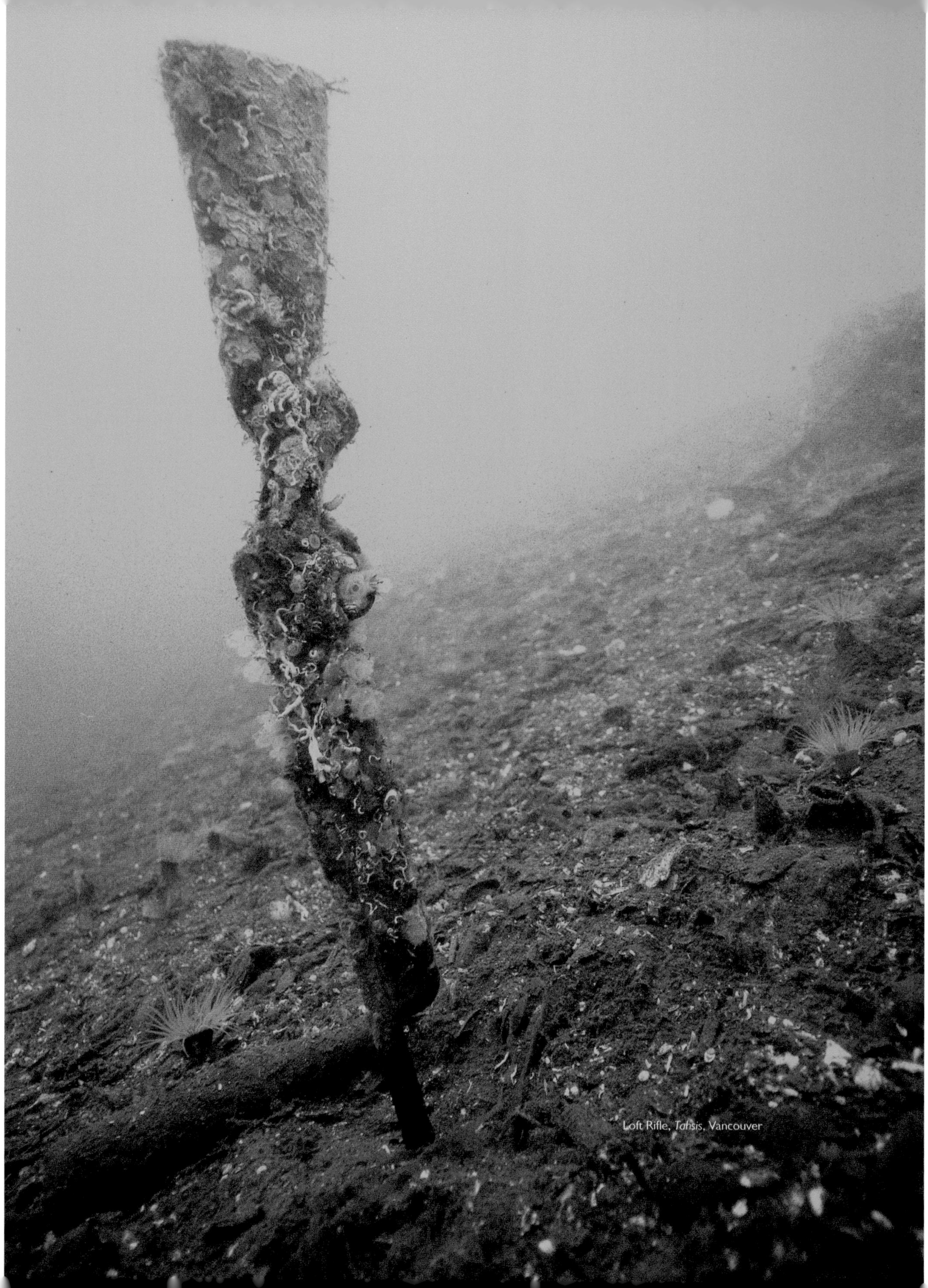

Loft Rifle, *Tahsis*, Vancouver

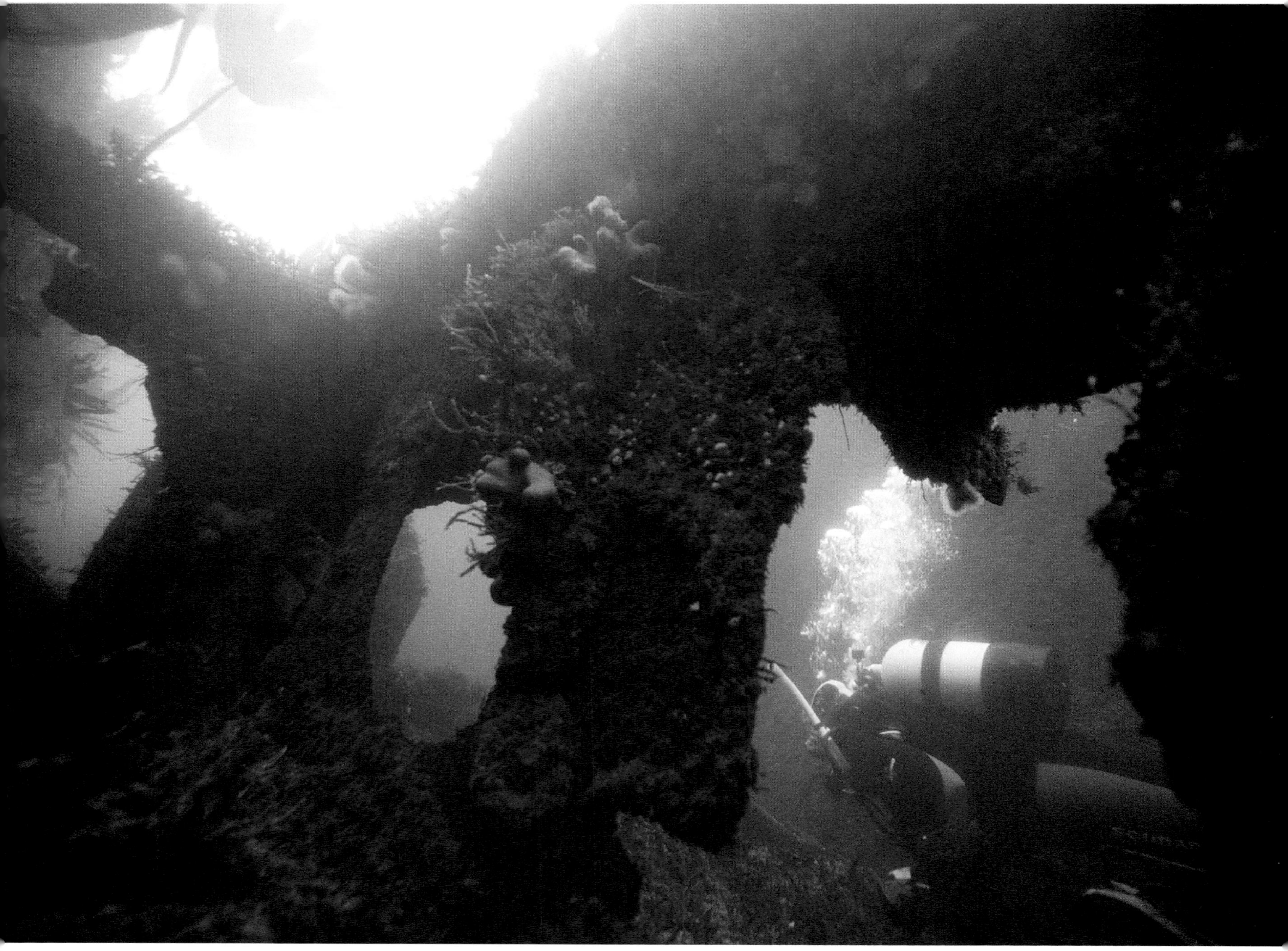

S.S. Alondra, The Kedges, Cork

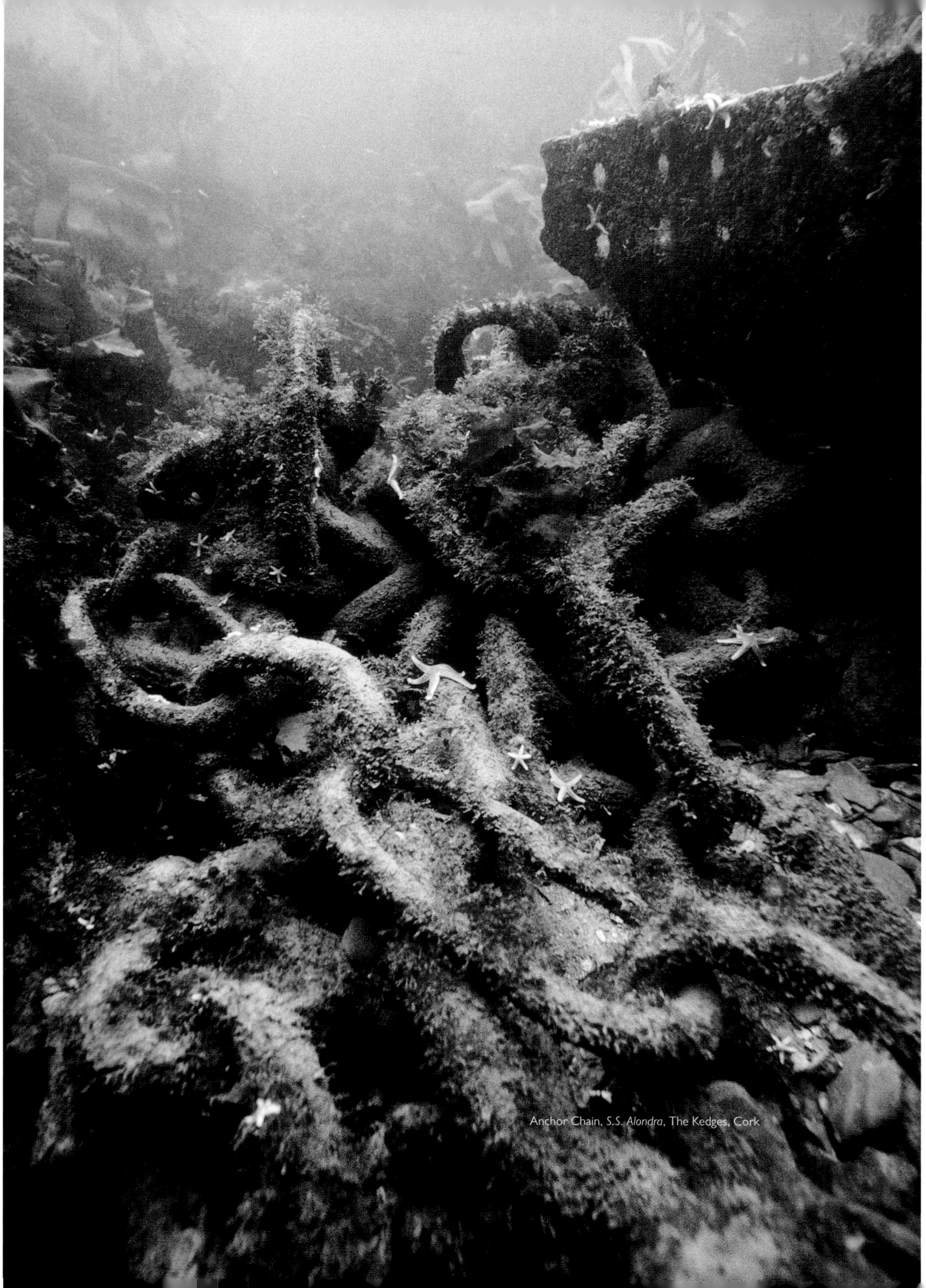

Anchor Chain, *S.S. Alondra*, The Kedges, Cork

It is not an easy job to find a sunken ship clearly marked on the chart by a line and a cross-bearing. In many cases only the local fishermen were able to take us to the spot with any certainty, and then only after long palavers over a glass of anisette.

Captain Philippe Tailliez, *To Hidden Depths*, 1954

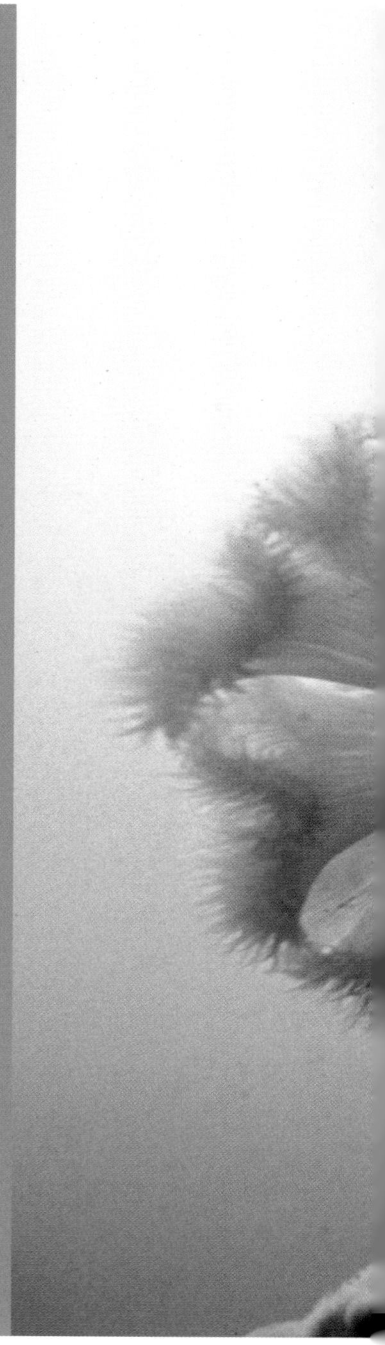

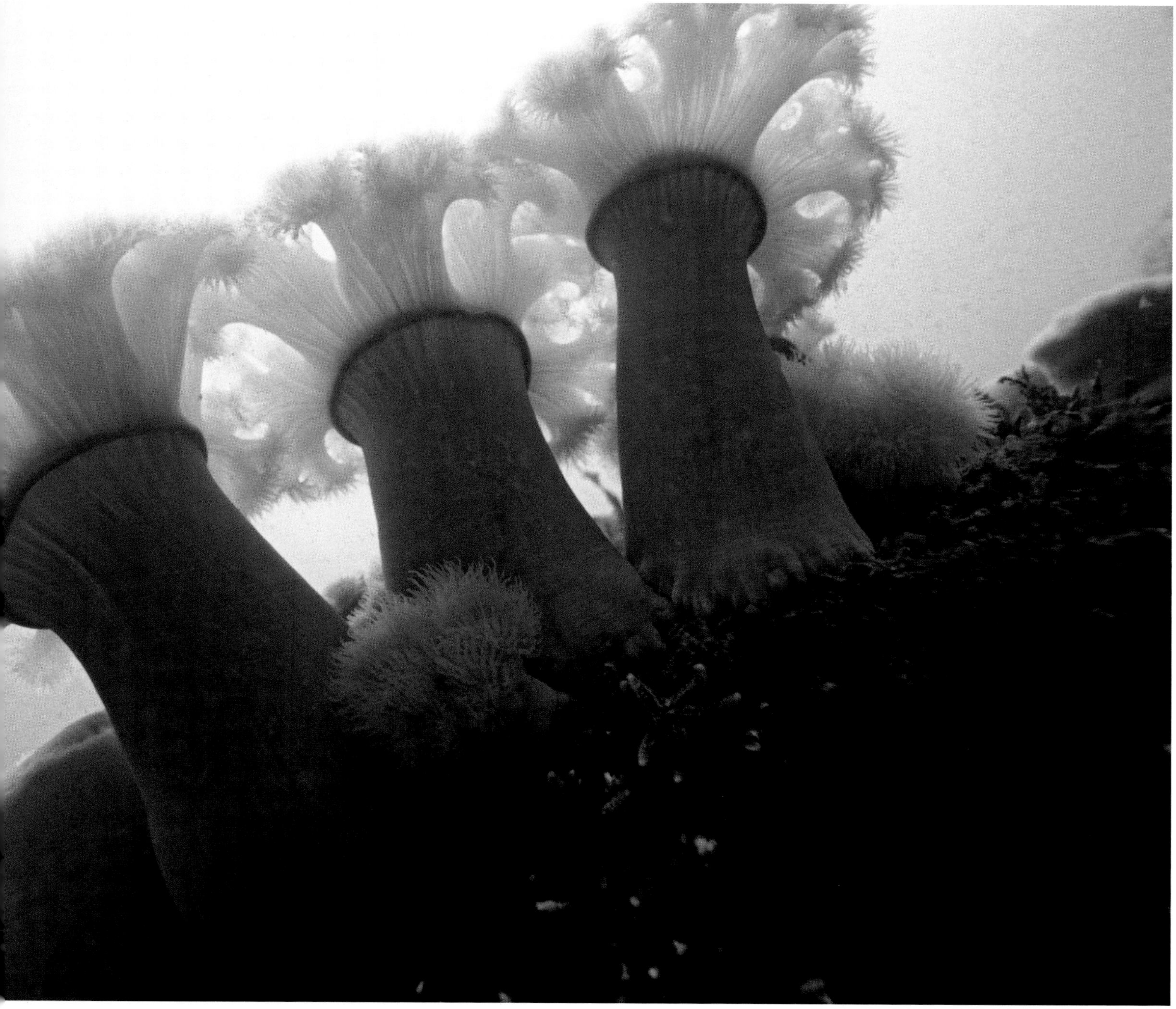

Plumose Anemones, Cape Clear, Cork

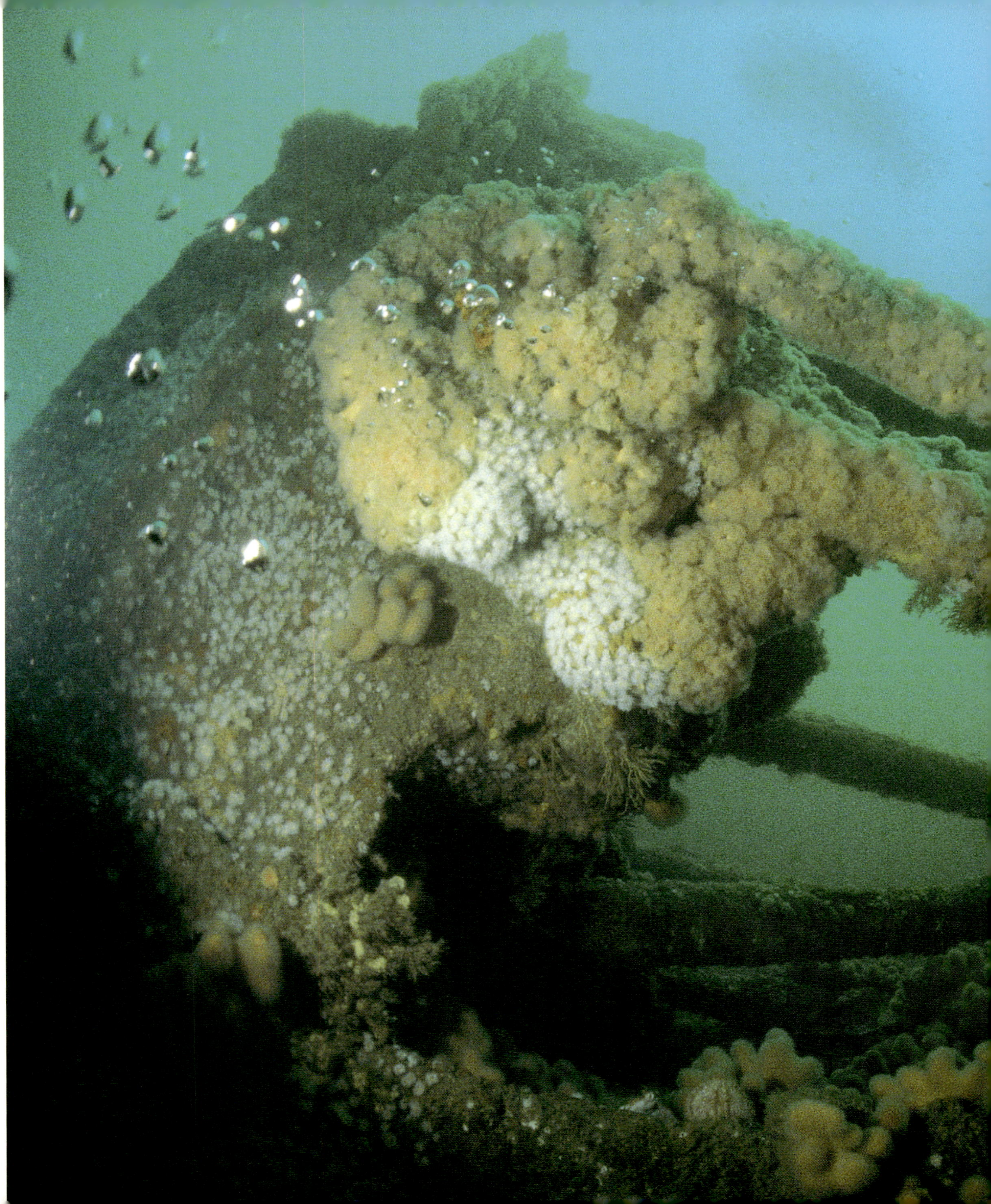

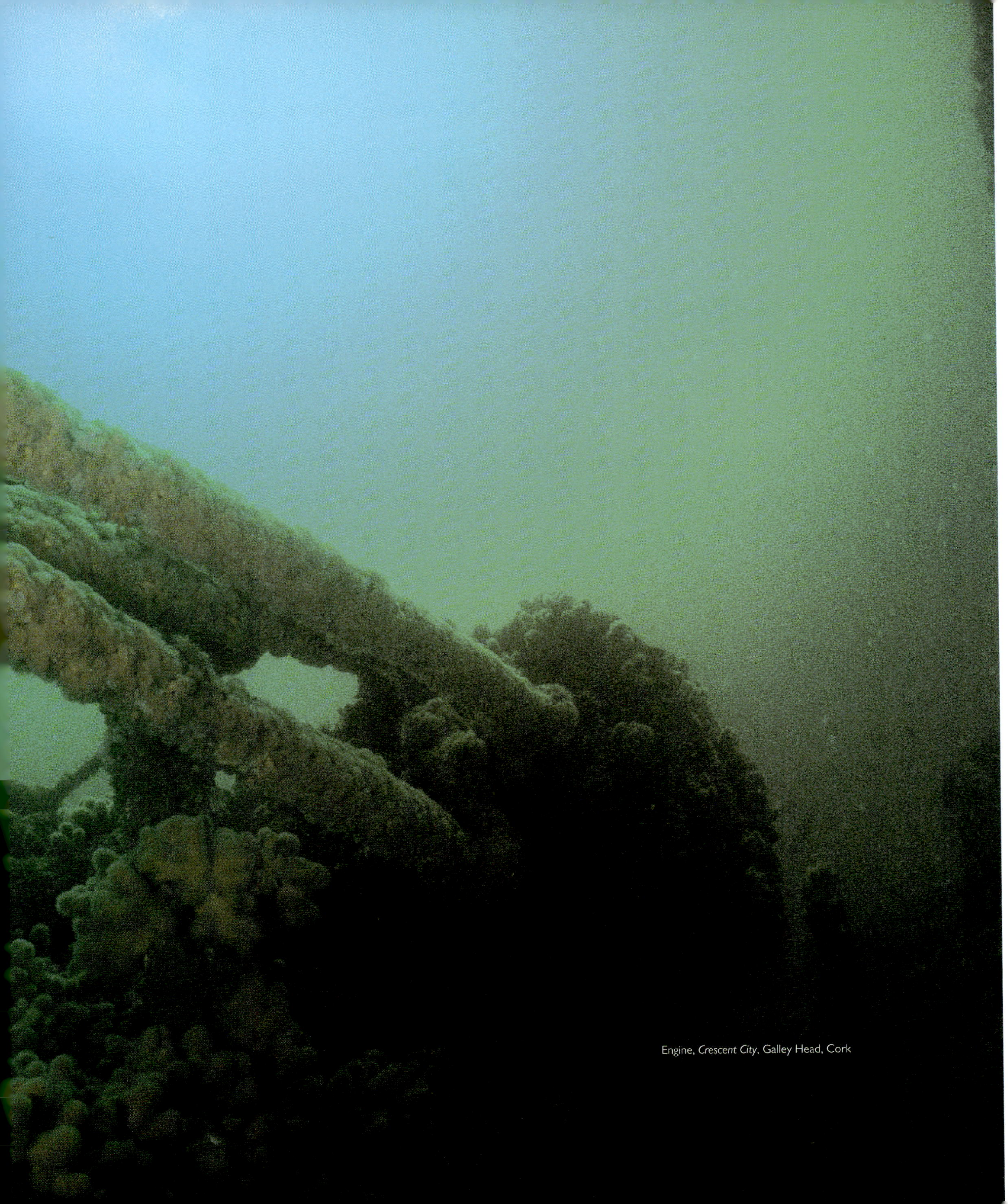

Engine, *Crescent City*, Galley Head, Cork

Methought I saw a thousand fearful wrecks,
Ten thousand men that fishes gnawed upon,
Wedges of gold, great anchors, heaps of pearl,
Inestimable stones, unvalued jewels,
All scattered in the bottom of the sea.

William Shakespeare, *King Richard III*, 1597

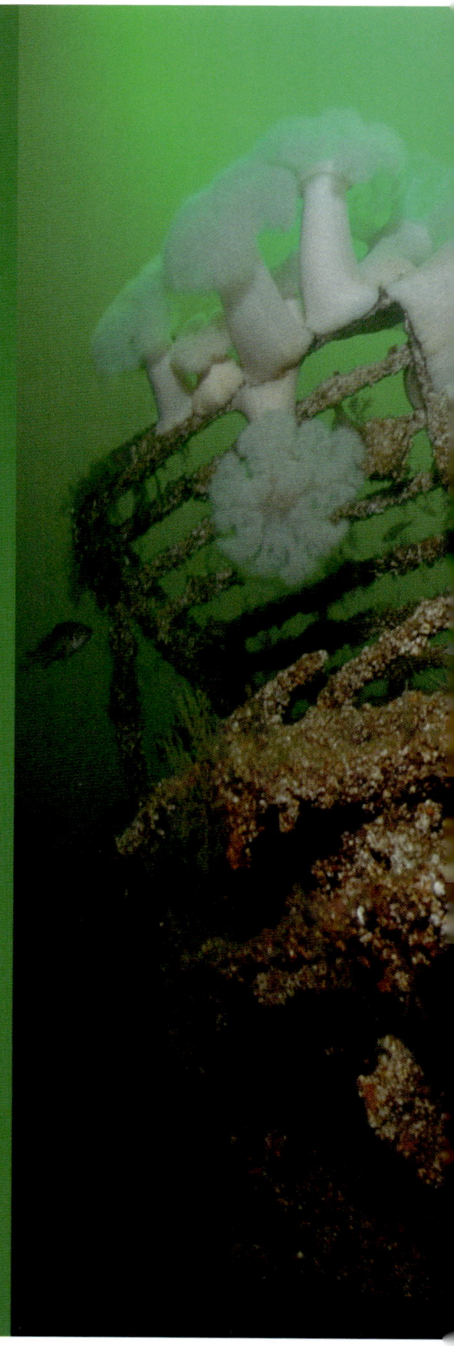

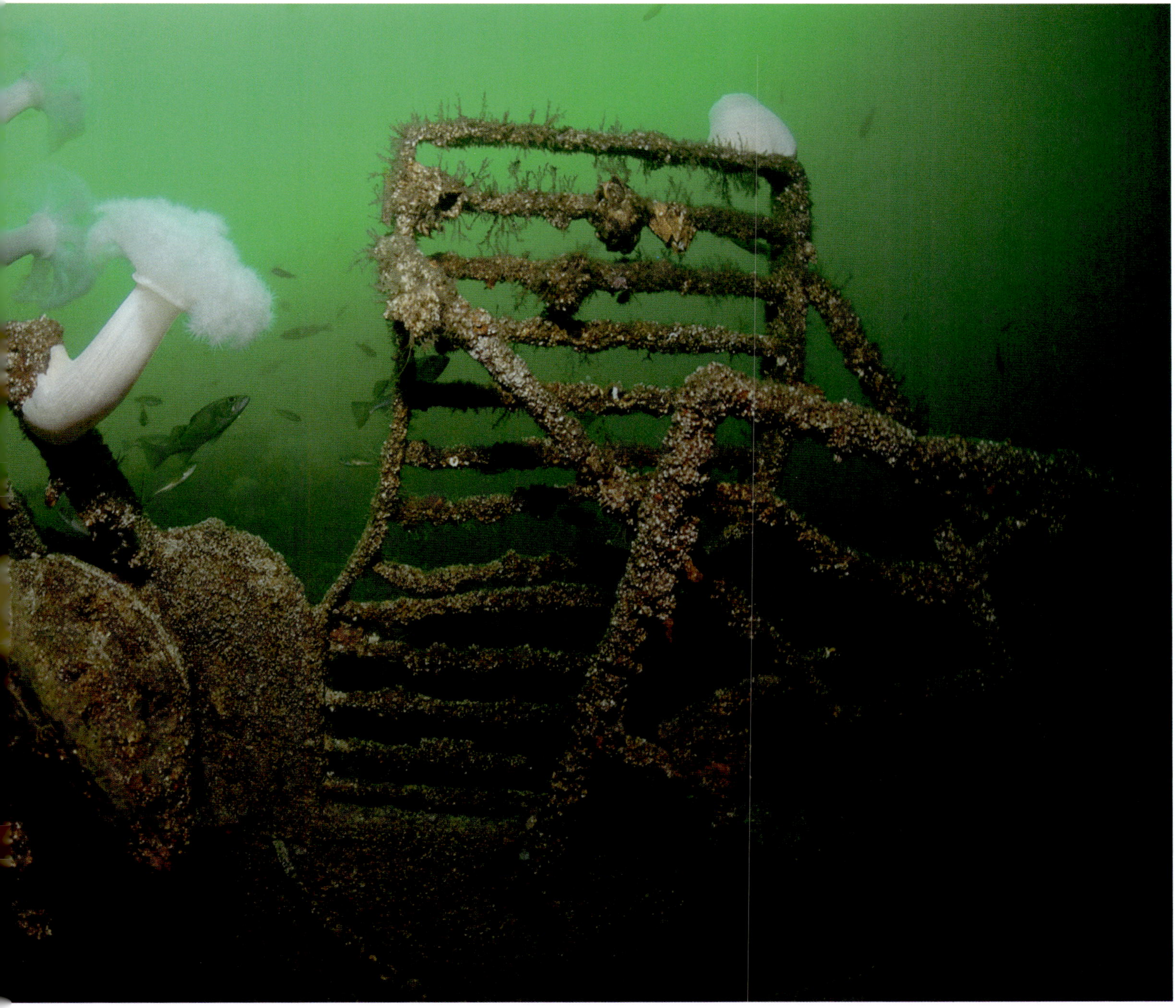

Vanlene, Barkley Sound, Vancouver

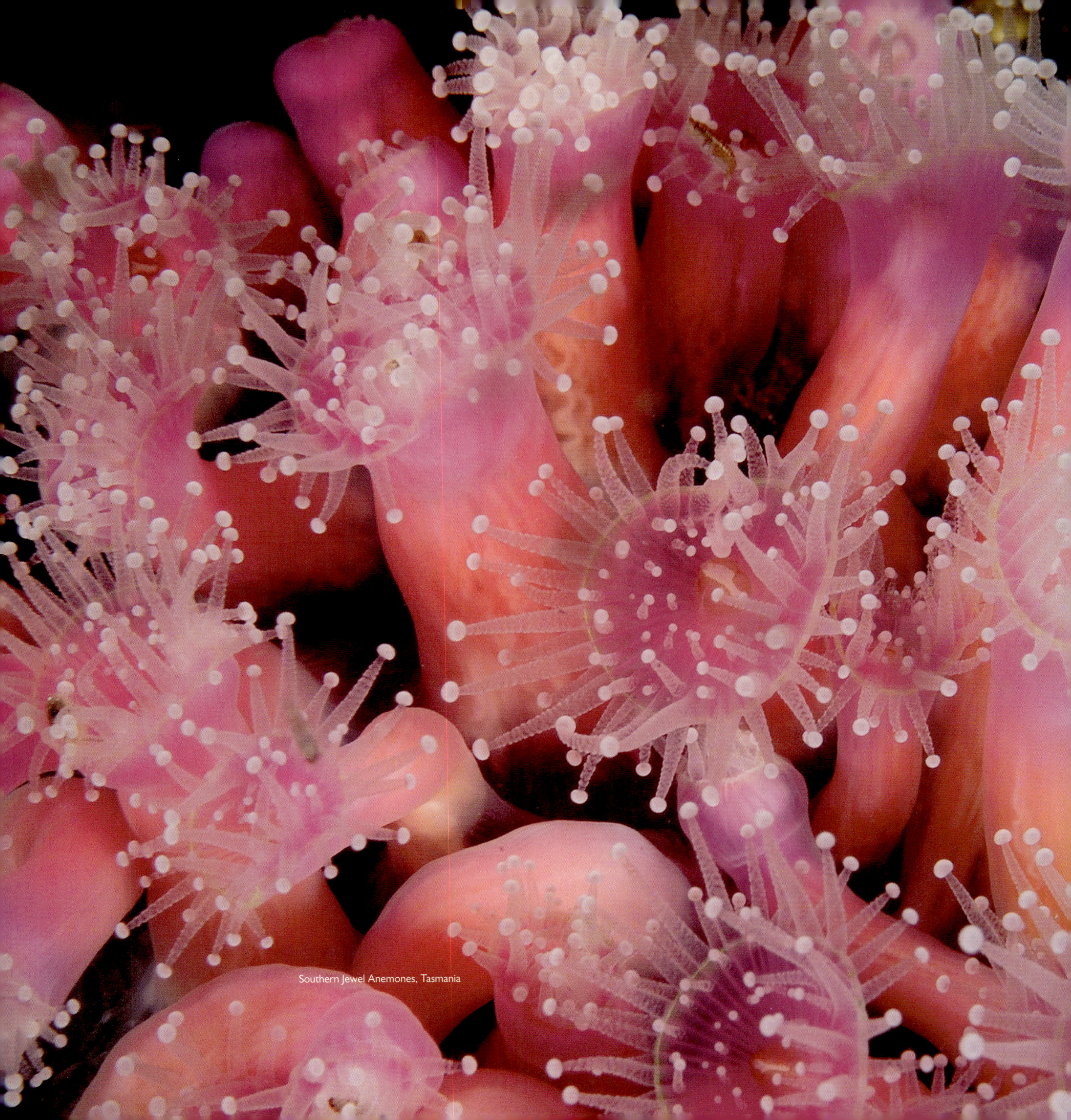

Southern Jewel Anemones, Tasmania

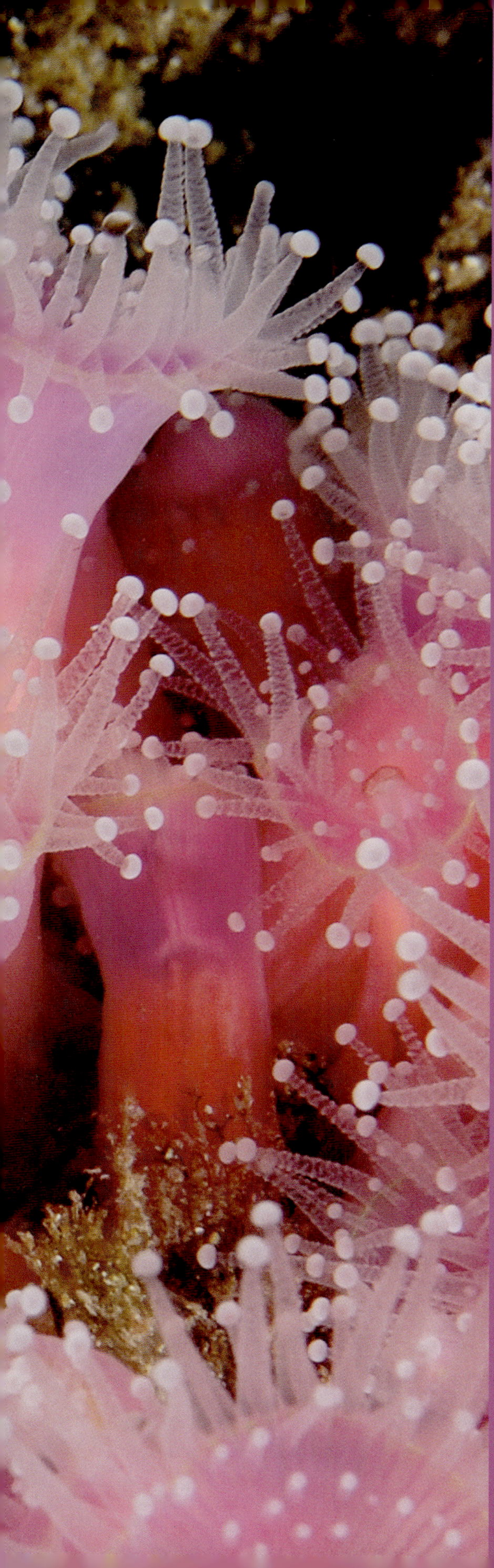

SEA DREAMS

The experience of swimming freely underwater is an intensely personal one. The sights and feelings of simply being in this underwater world cannot be adequately described in land-world terms. It is so outside our everyday lives that we simply have not yet developed a suitable vocabulary.

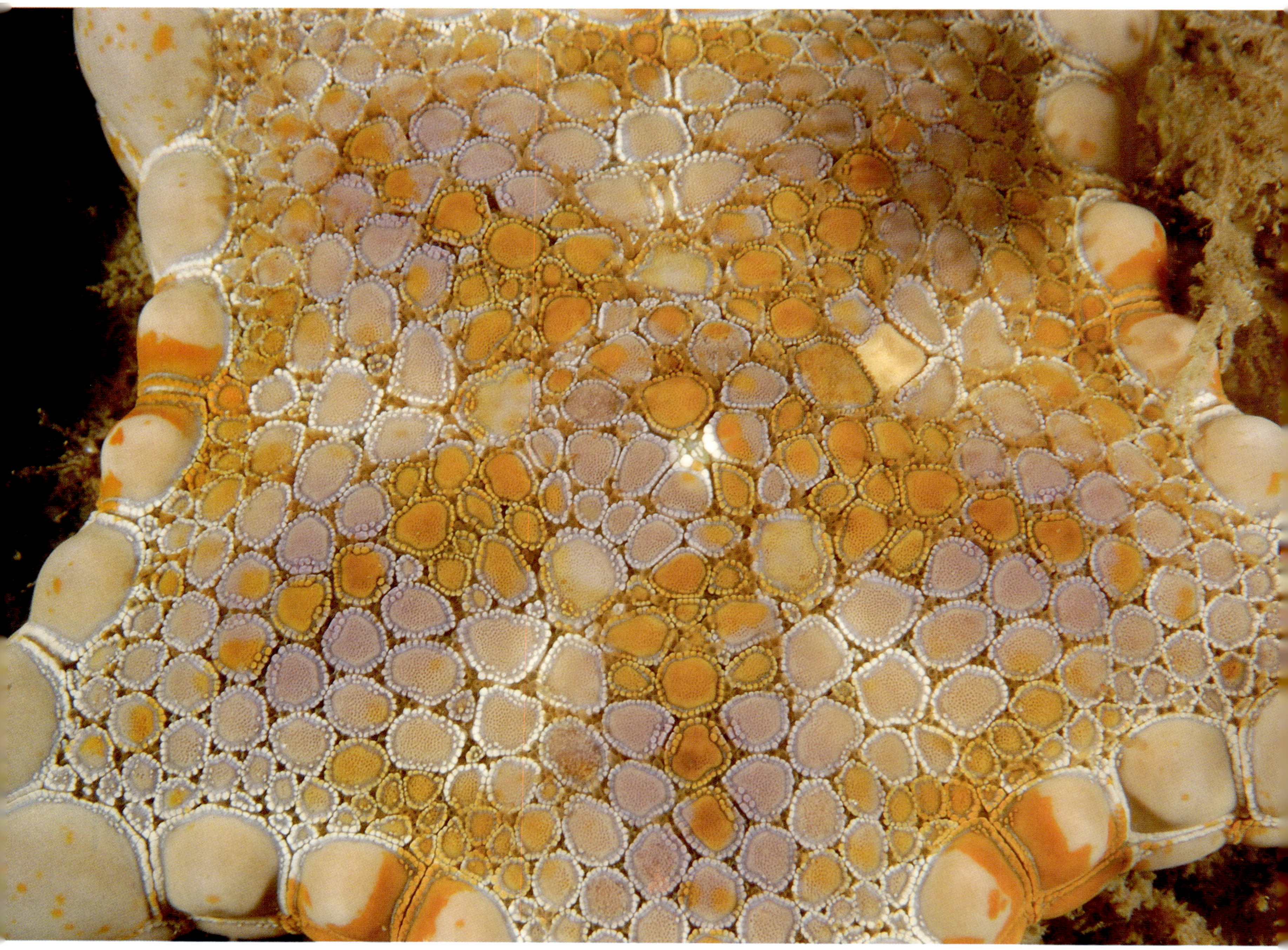

Biscuit Star, Waterfall Bay, Tasmania

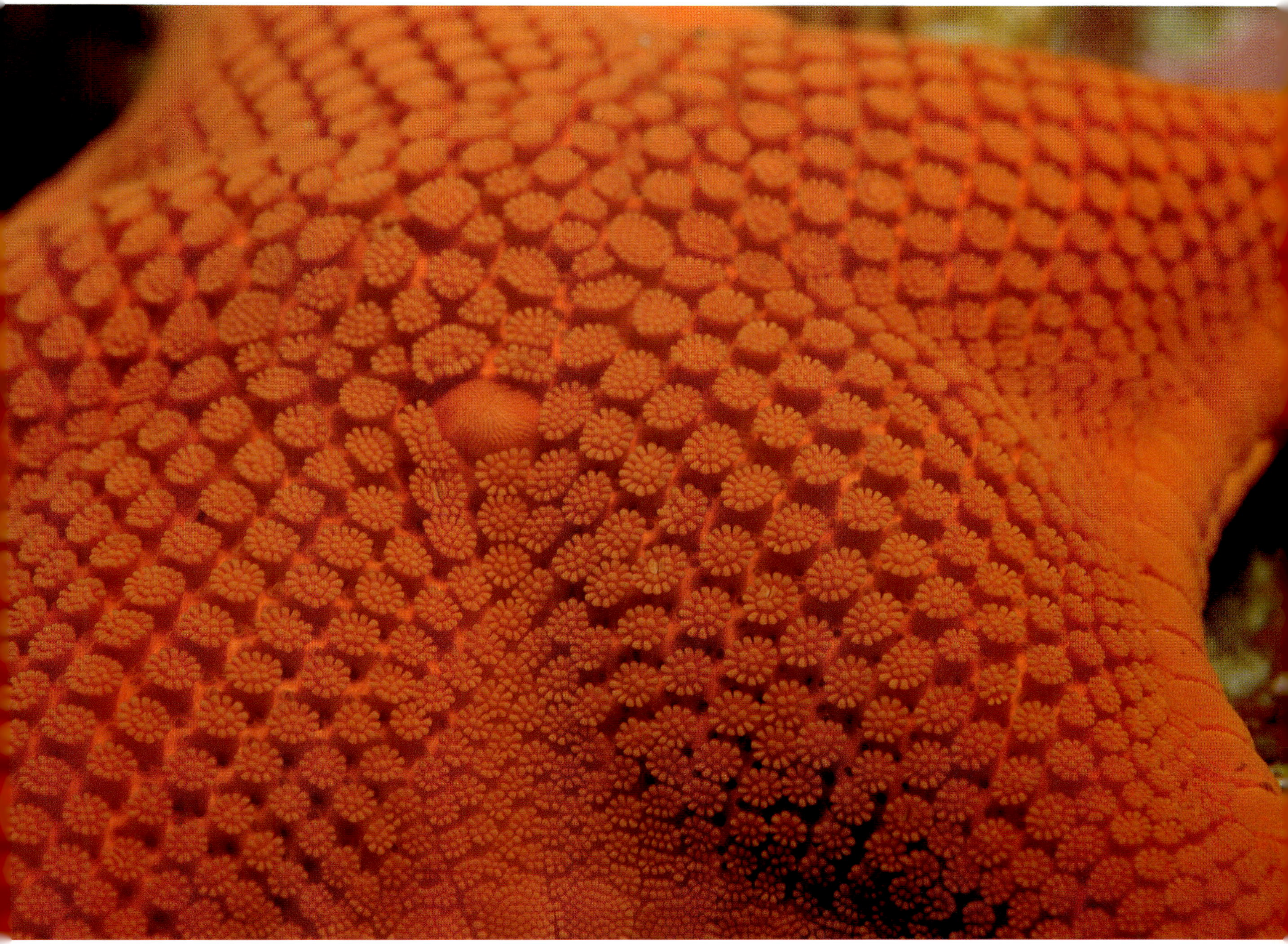

Vermilion Sea Star, Quatsino Sound, Vancouver

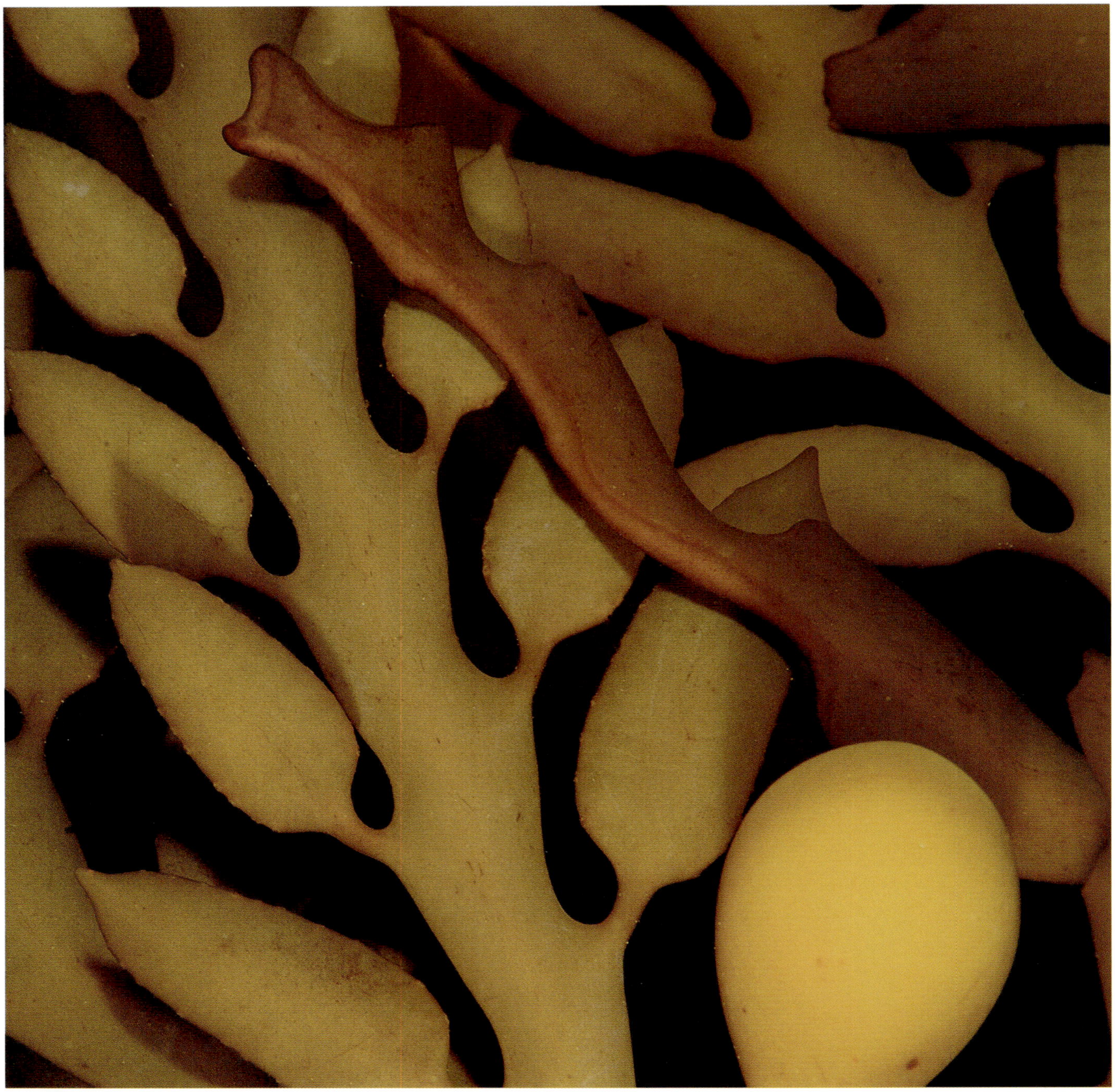

Kelp Detail, Waterfall Bay, Tasmania

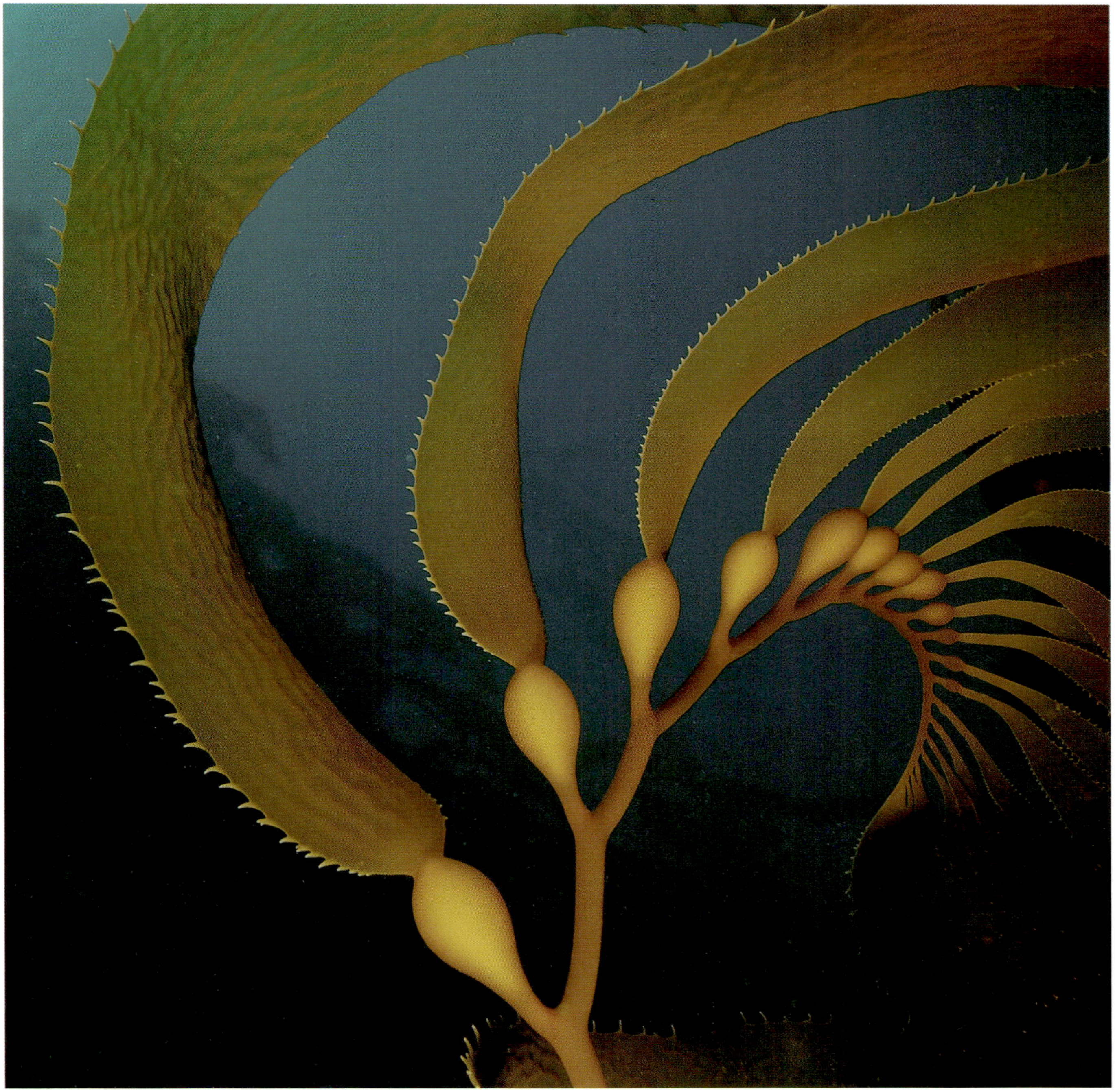

Giant Kelp, Fortescue Bay, Tasmania

You are standing on a metal ladder in water up to your neck. Something round and heavy is slipped gently over your head, and a metal helmet rests on your shoulders. Thus were the knights of old helmed by their squires for the grim business of war. Instead of the slotted visor, however, you find two large frames of glass before your eyes. Turning your head you see emerald waves breaking upon the distant beach of ivory, backed by feathery palms waving in the sunlight against a sky of pure azure.

You wave good-by to your grinning friend at the pump, and slowly descend, climbing down step by step. For a brief space of time the palms and the beach show intermittently through waves which are now breaking over your very face. Then the world changes. There is no more harsh sunlight but delicate blue-greens with a fluttering of shadows everywhere.

William Beebe, *Beneath Tropic Seas*, 1928

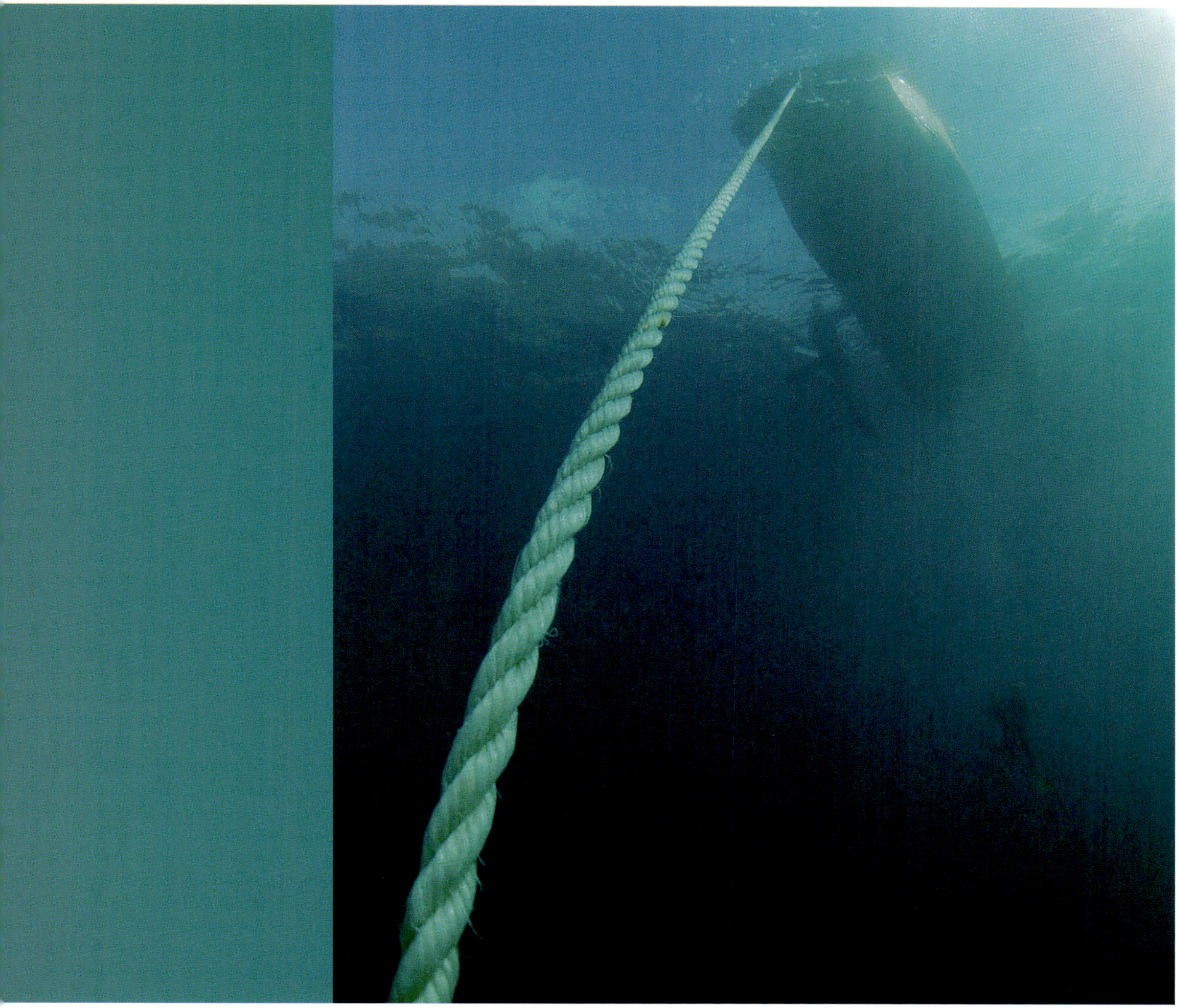

Descent, Munro Bight, Tasmania

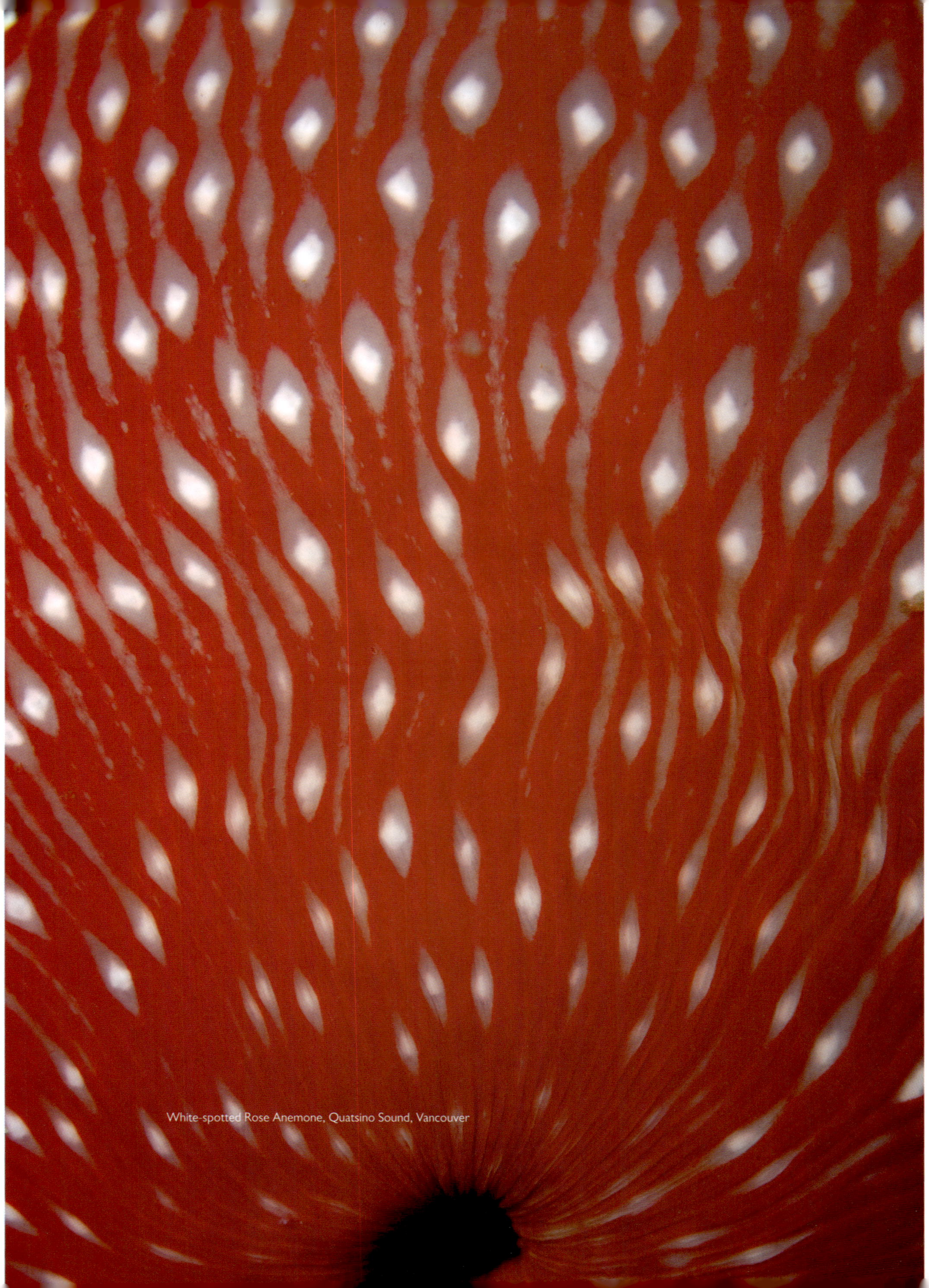

White-spotted Rose Anemone, Quatsino Sound, Vancouver

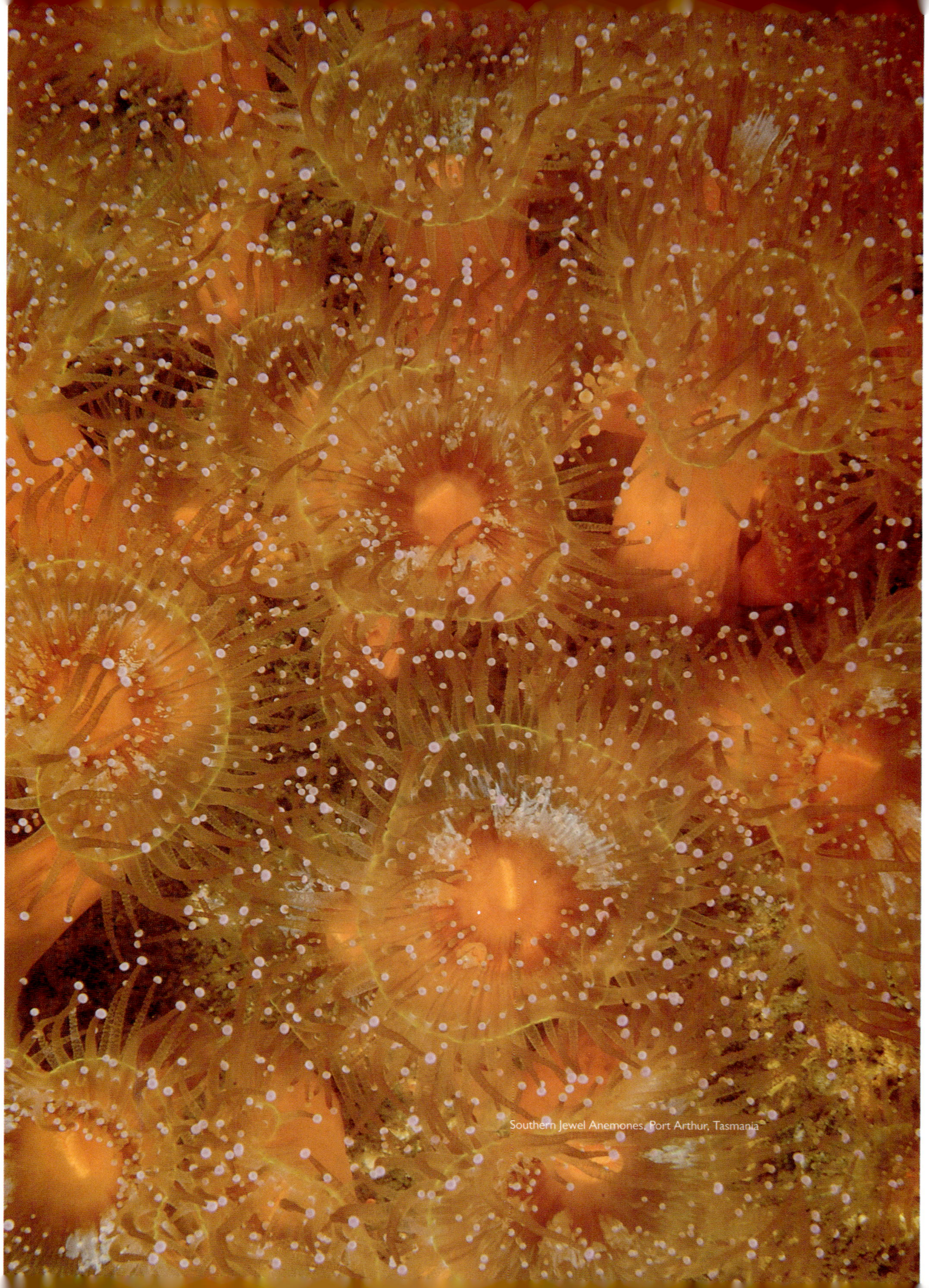

Southern Jewel Anemones, Port Arthur, Tasmania

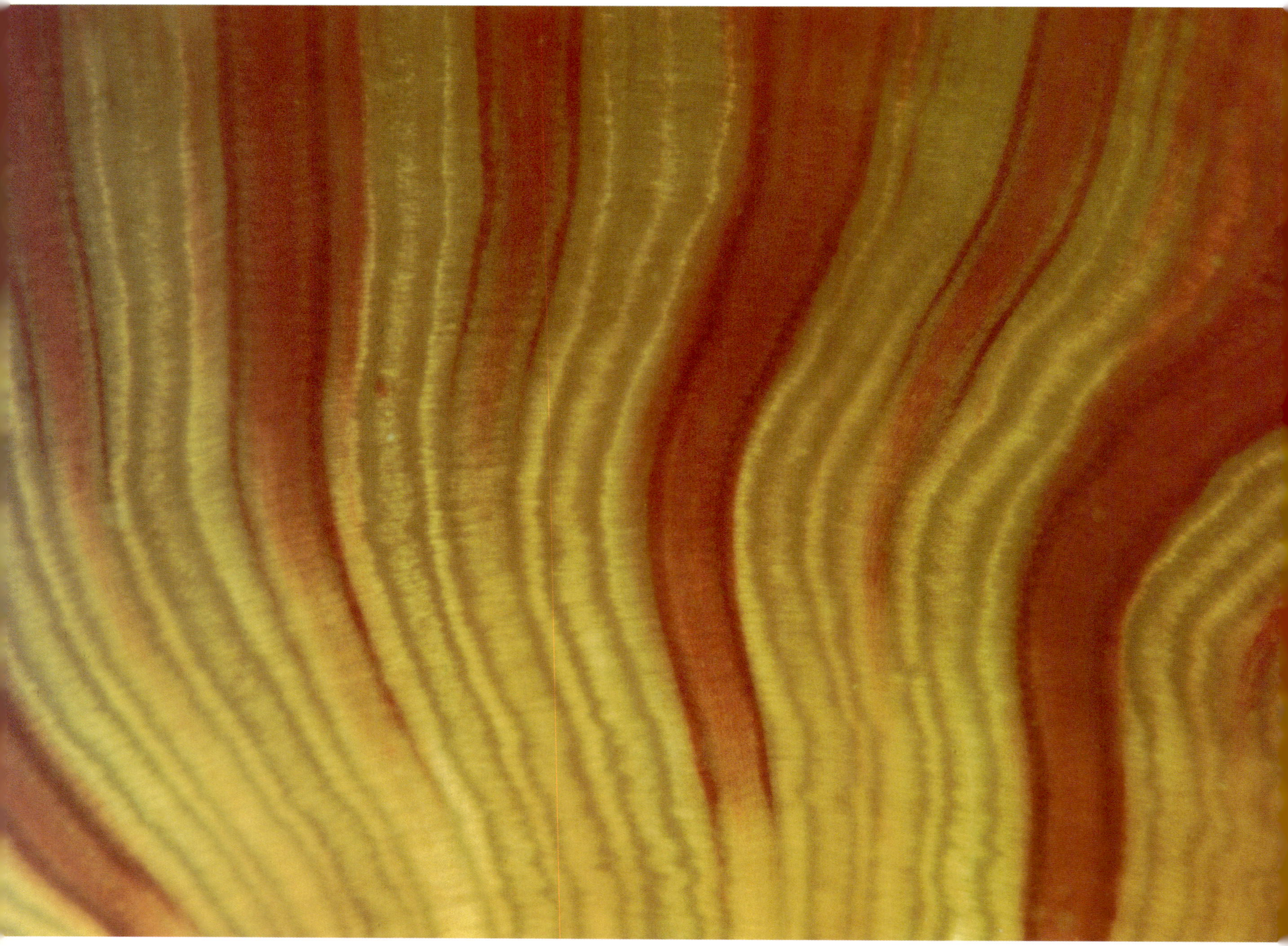

Anemone Detail, Quatsino Sound, Vancouver

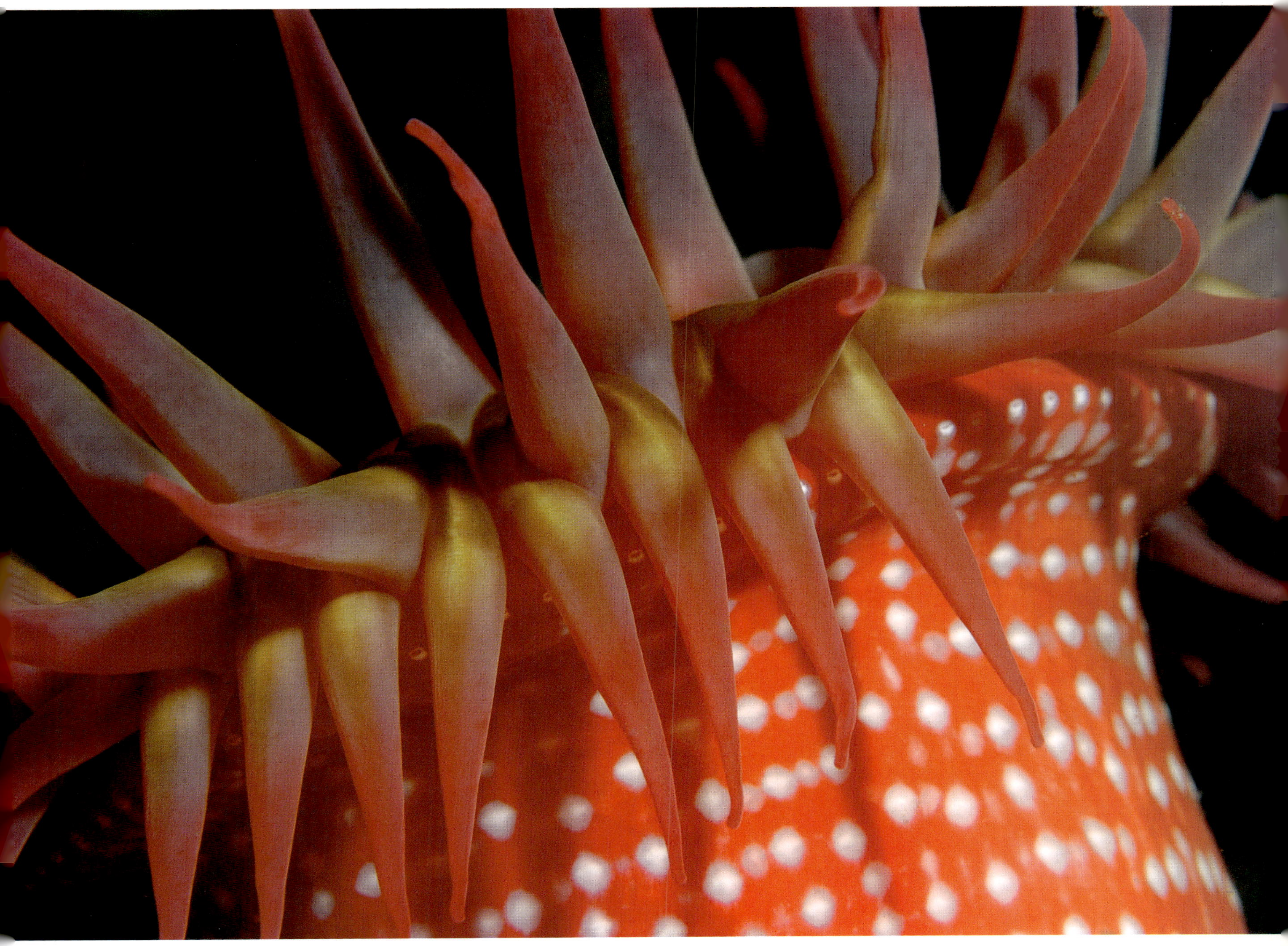

White-spotted Rose Anemone, Quatsino Sound, Vancouver

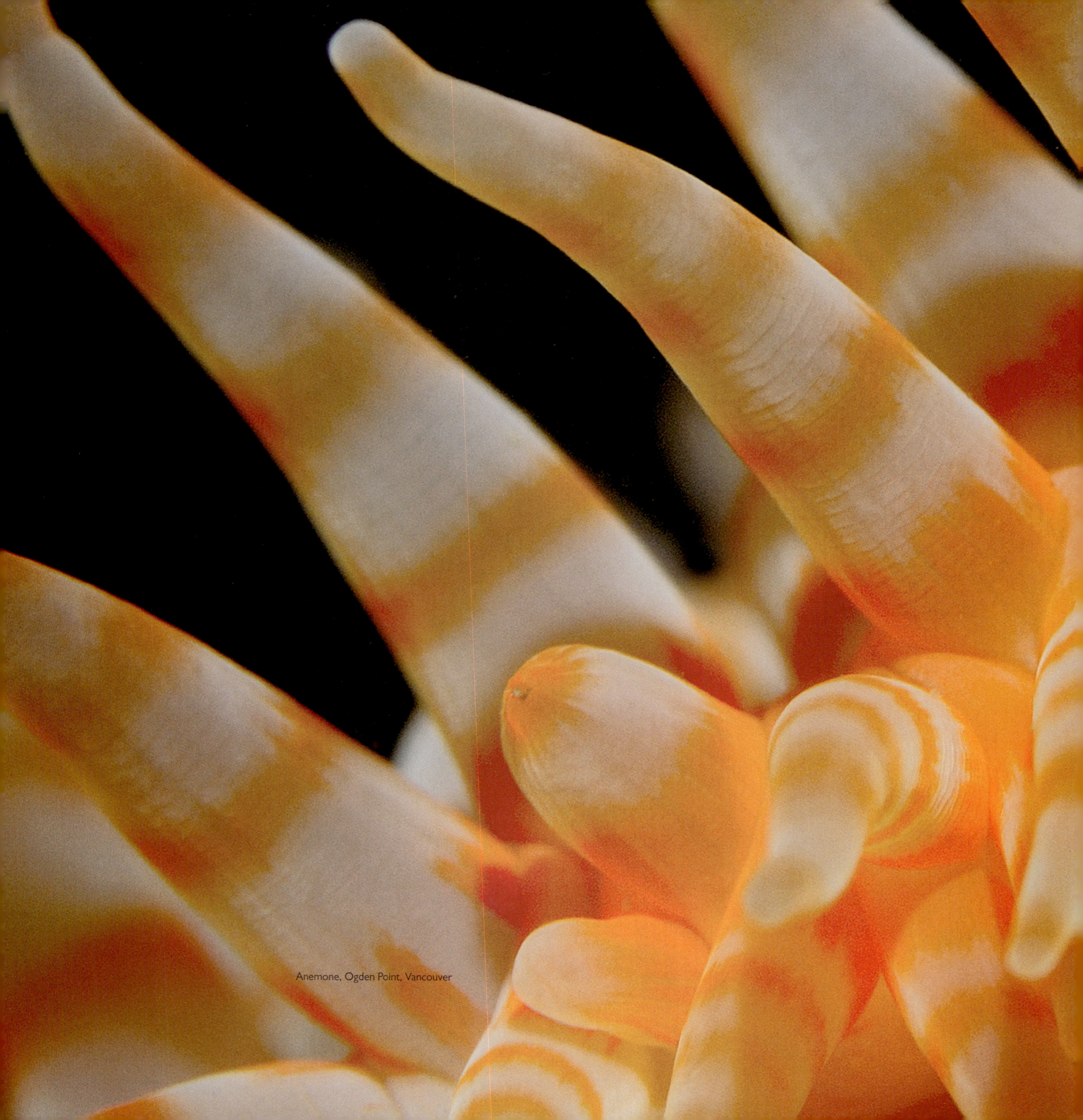

Anemone, Ogden Point, Vancouver

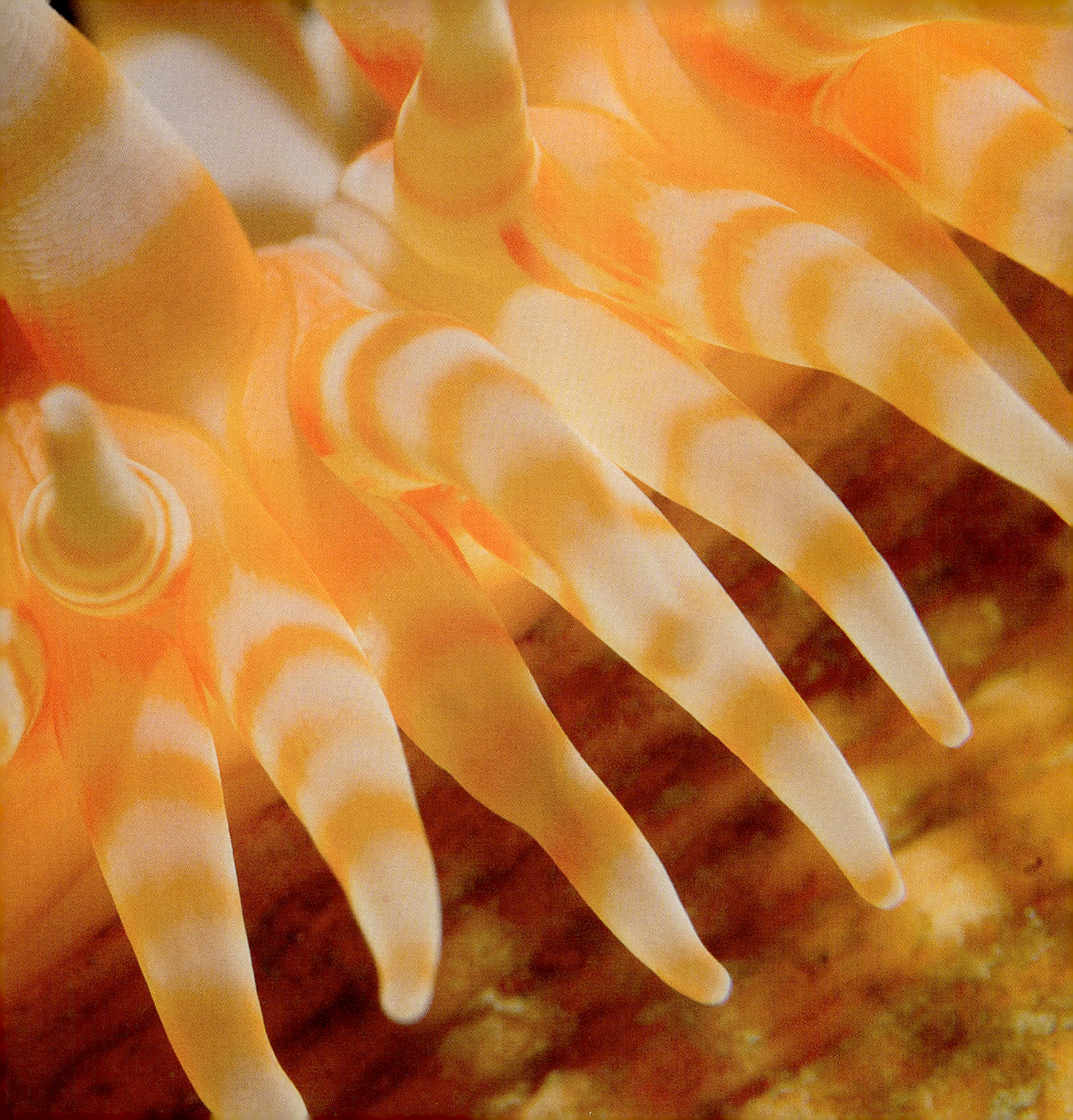

Here, I am a vagrant, the water so endlessly clear that it offers me no visible means of support. I often feel like I am falling, and my first impulse is to reach back and grab the gunwhale of the boat to anchor my senses. And then there is the silence. An infinite blue silence. I touch nothing, I smell nothing. I am completely alone. Even gravity is gone. In this moment I join a vast solitude. Taking a deep breath, I arch my body and dive, my sense of time and self recreated.

Christopher Newbert, *Within a Rainbowed Sea*, 1984

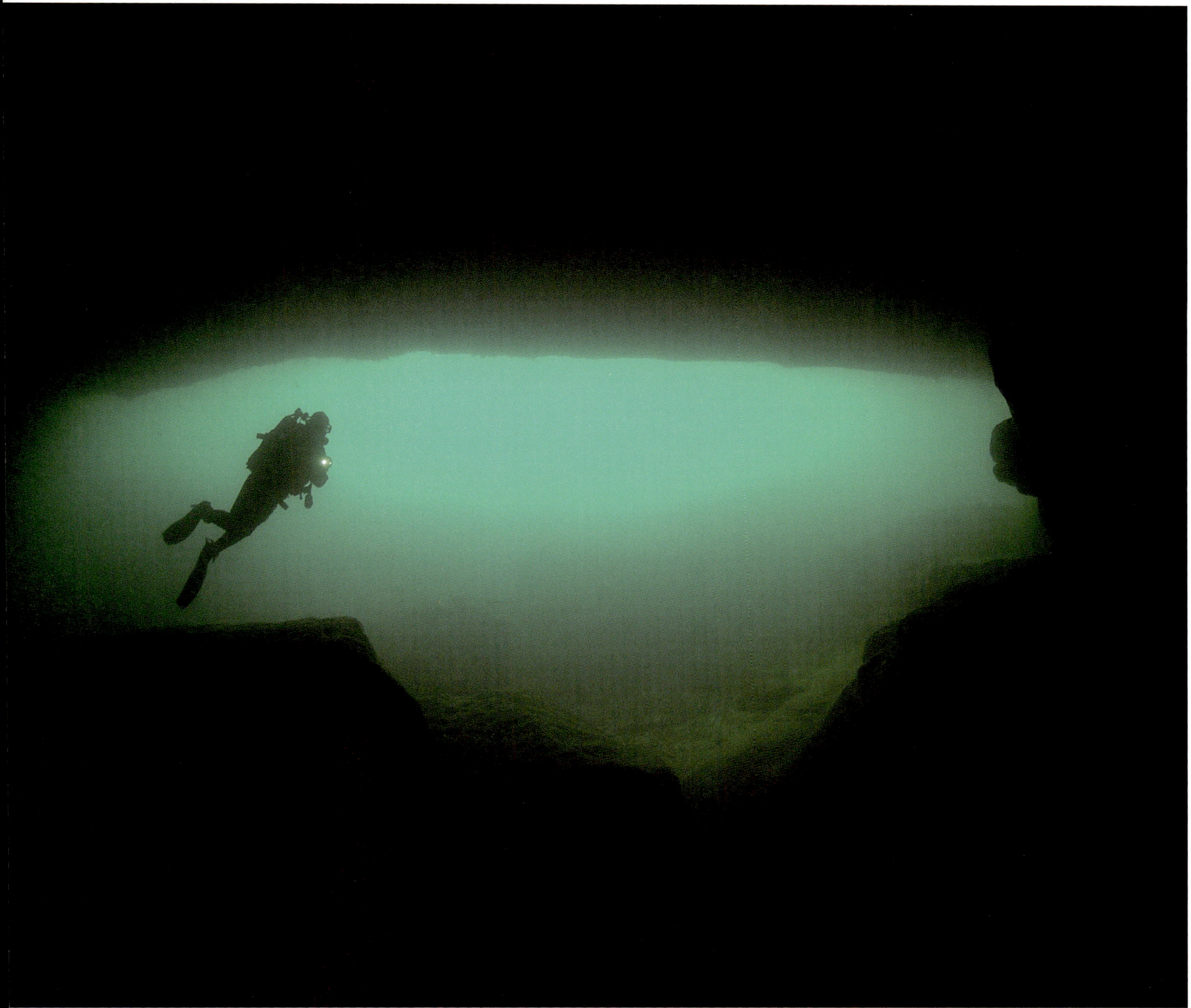

Worm Hole, Inishmore, Aran Islands

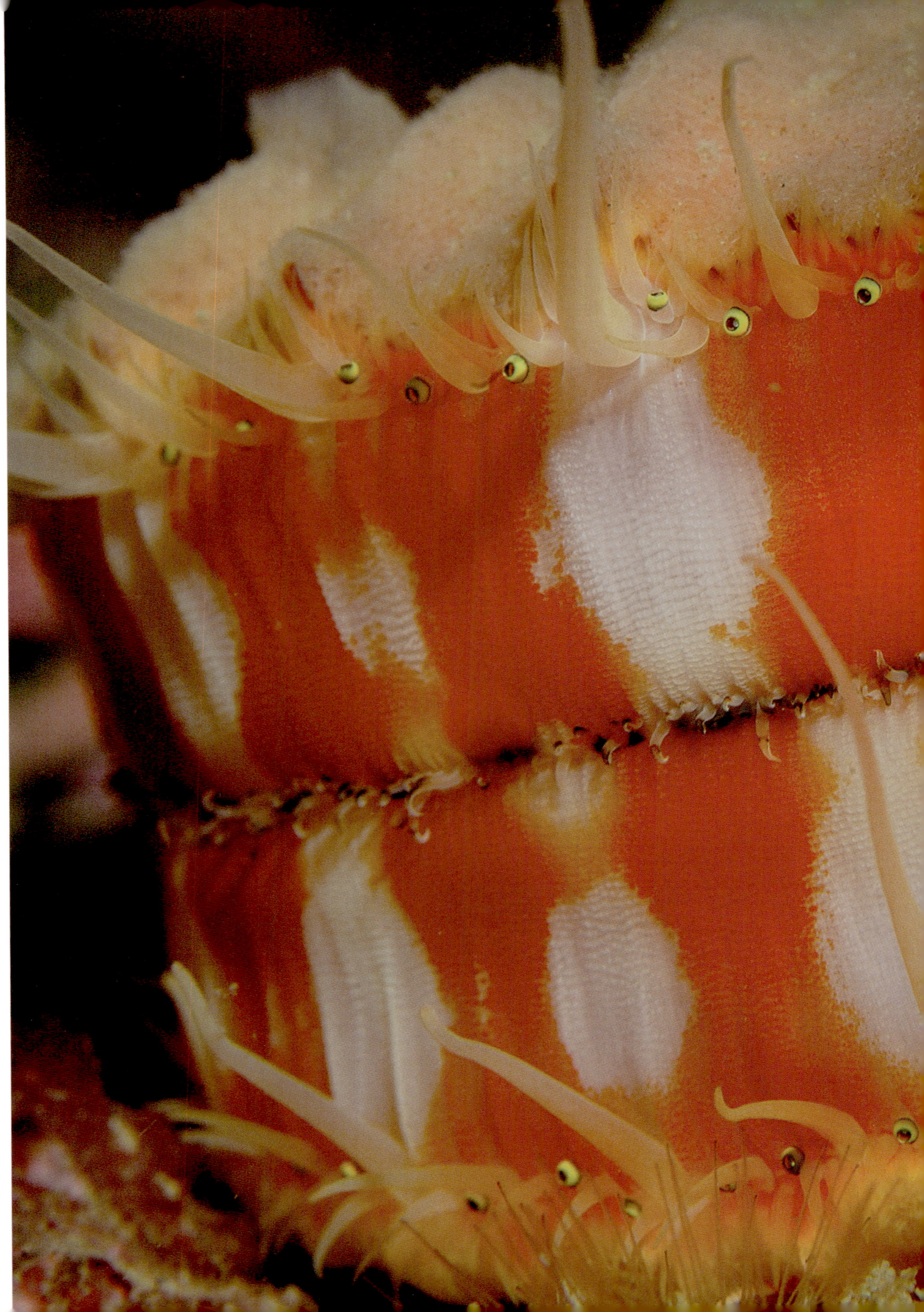

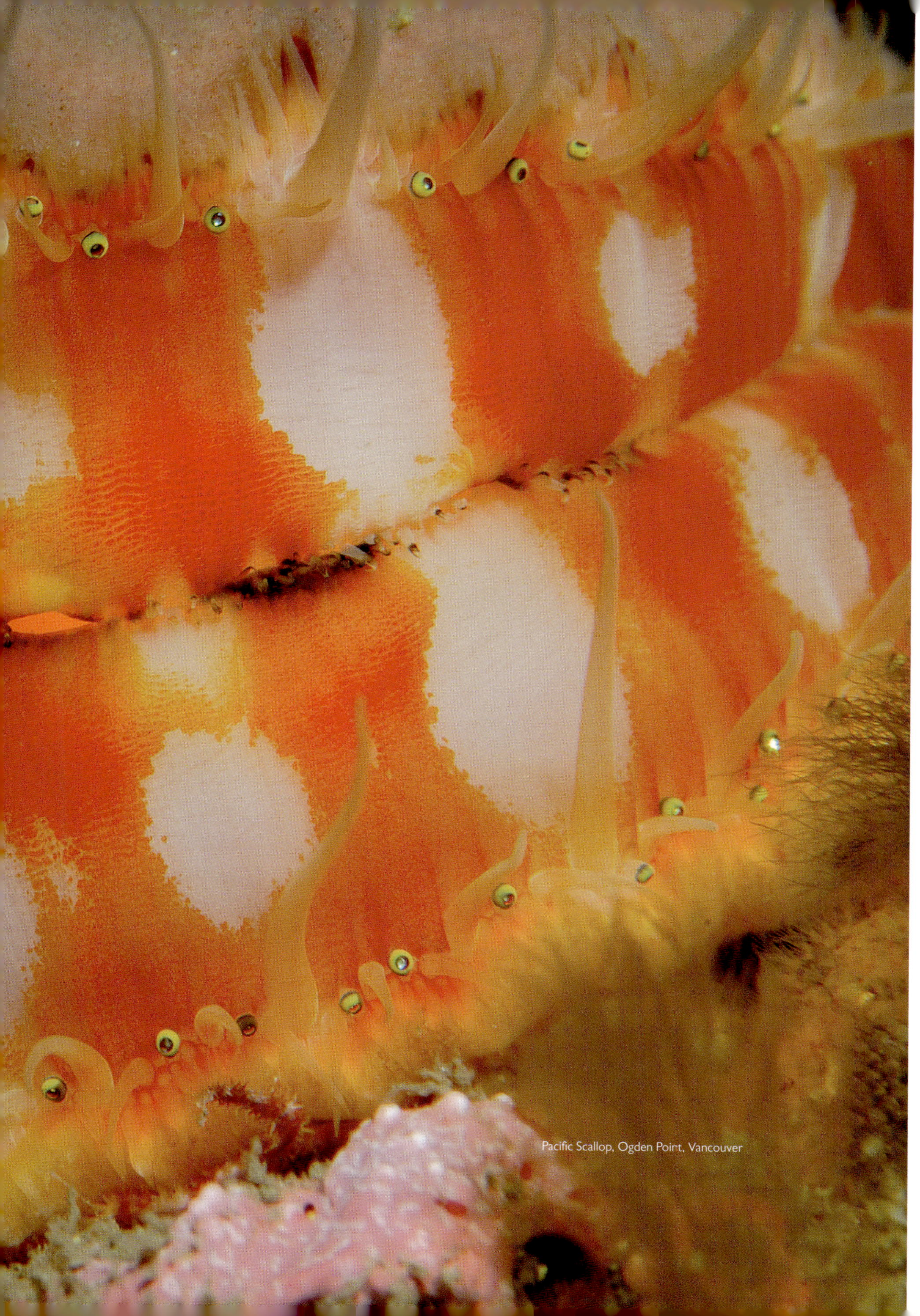

Pacific Scallop, Ogden Point, Vancouver

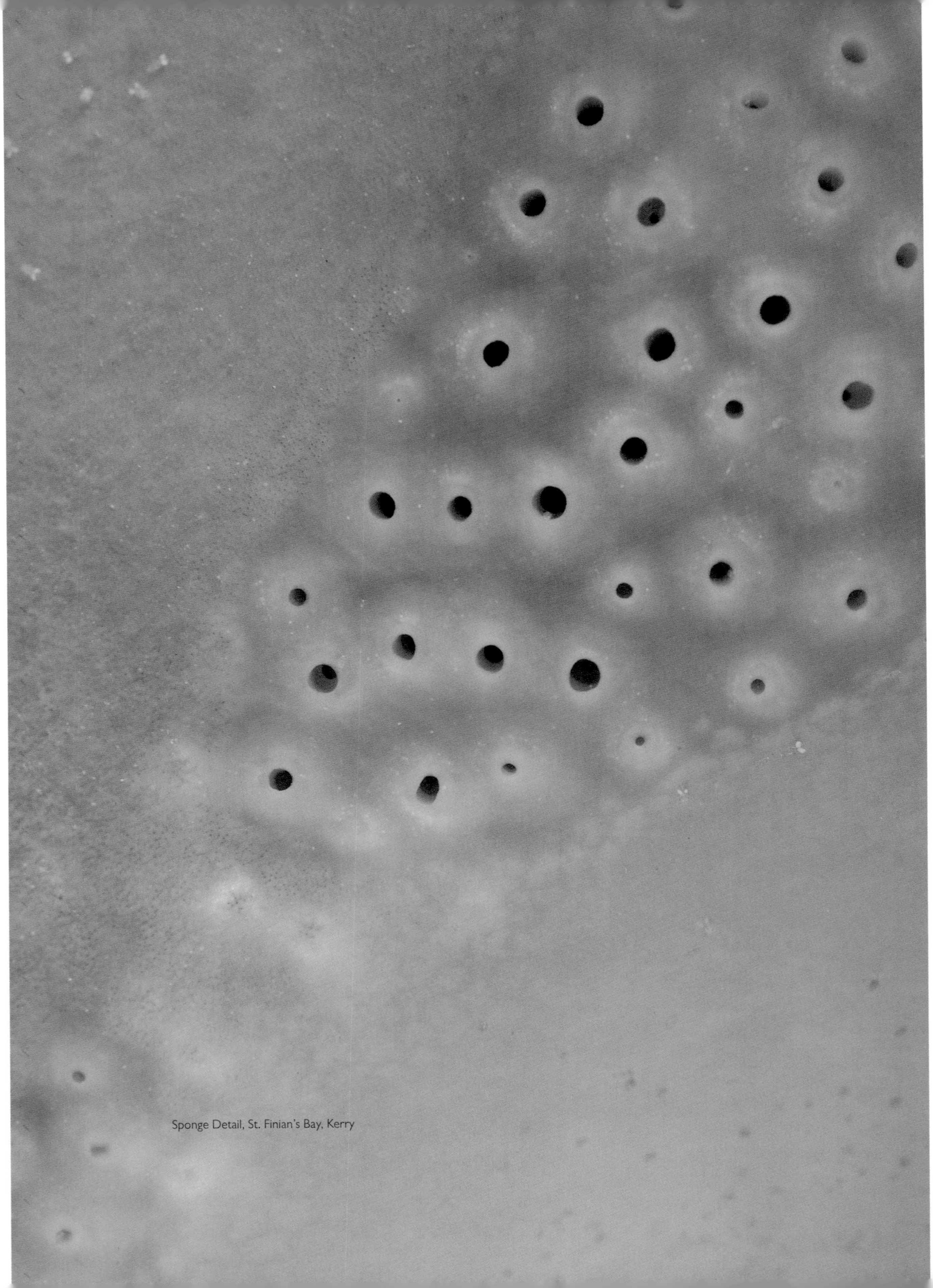

Sponge Detail, St. Finian's Bay, Kerry

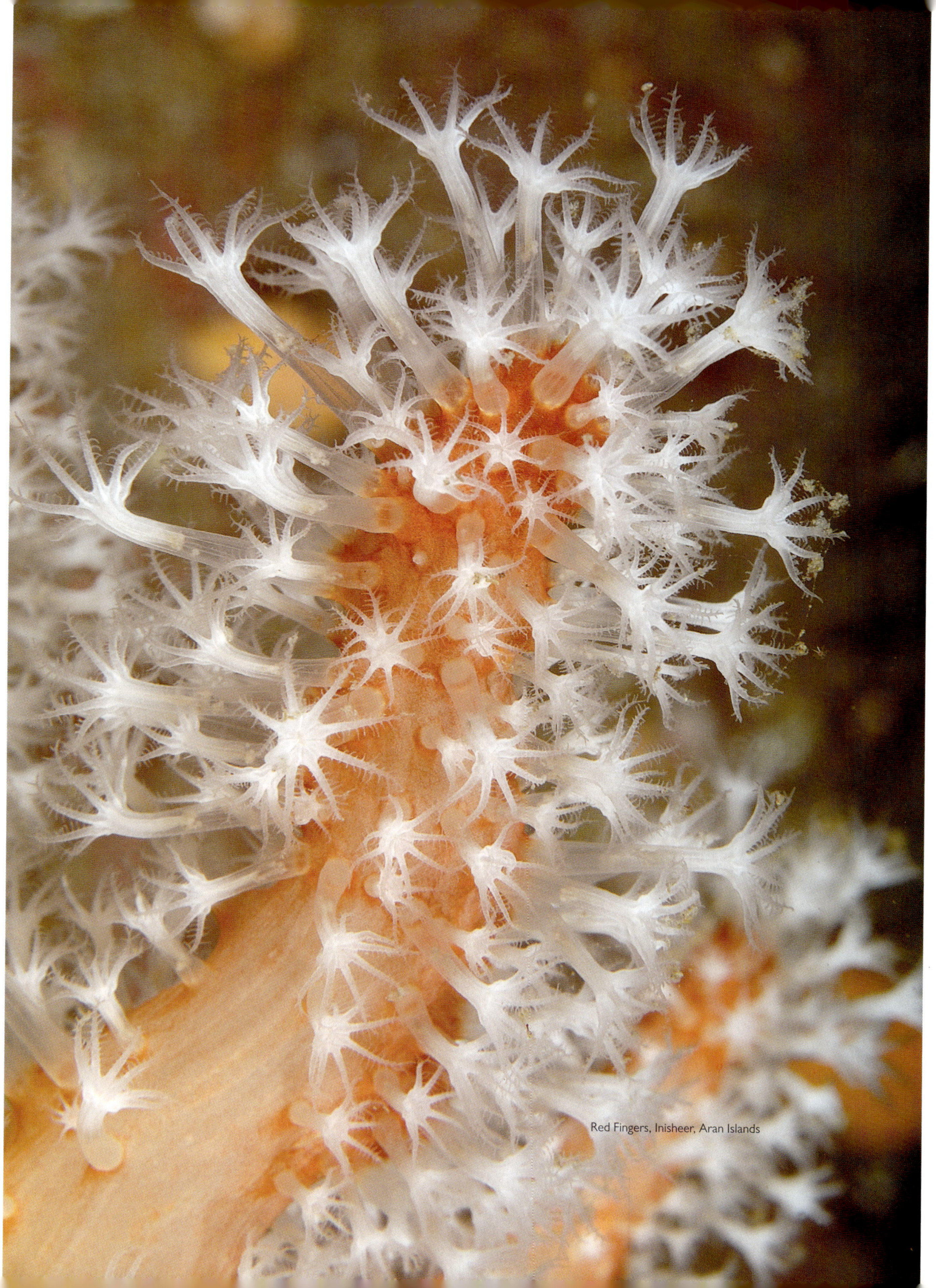

Red Fingers, Inisheer, Aran Islands

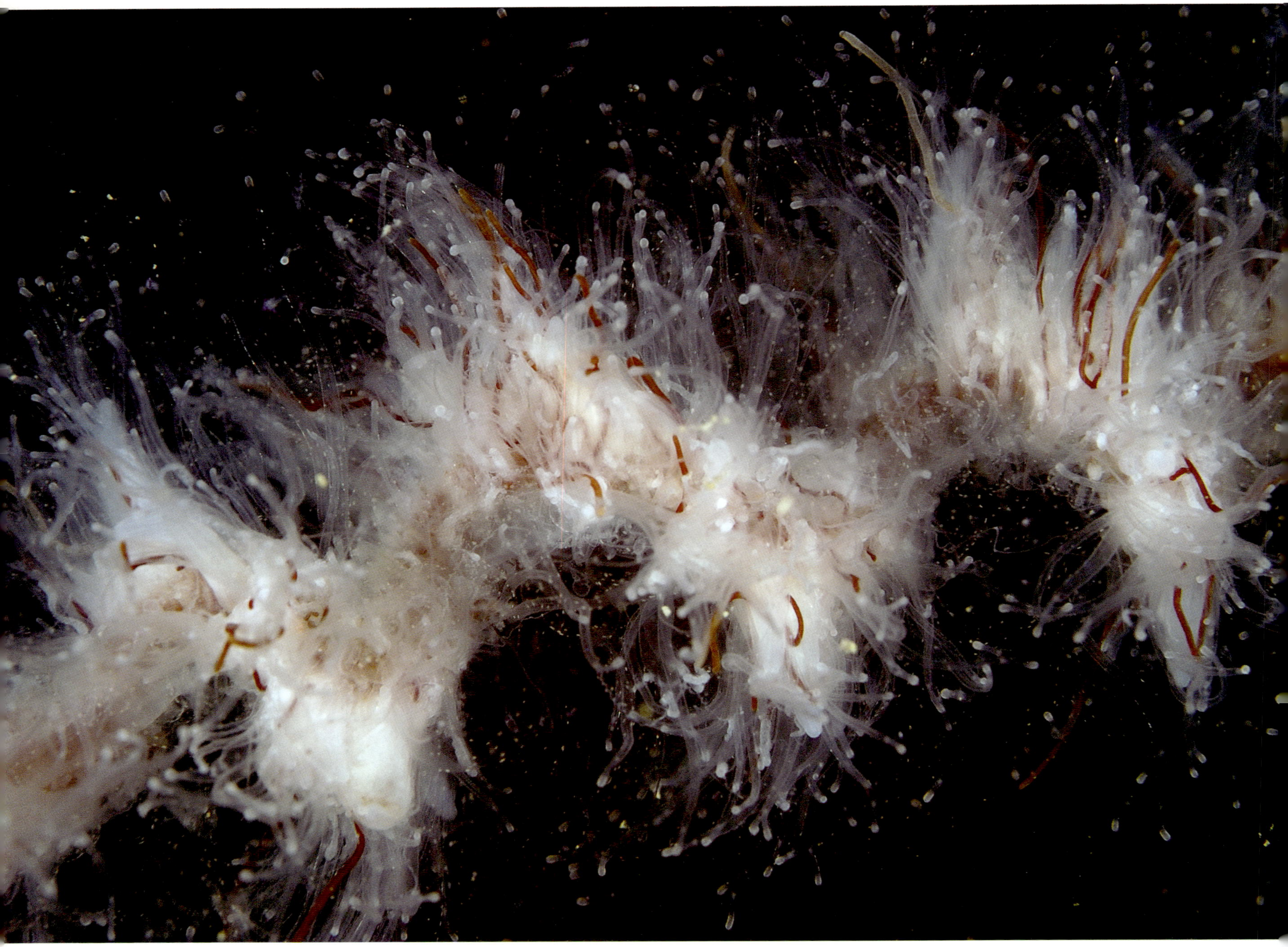

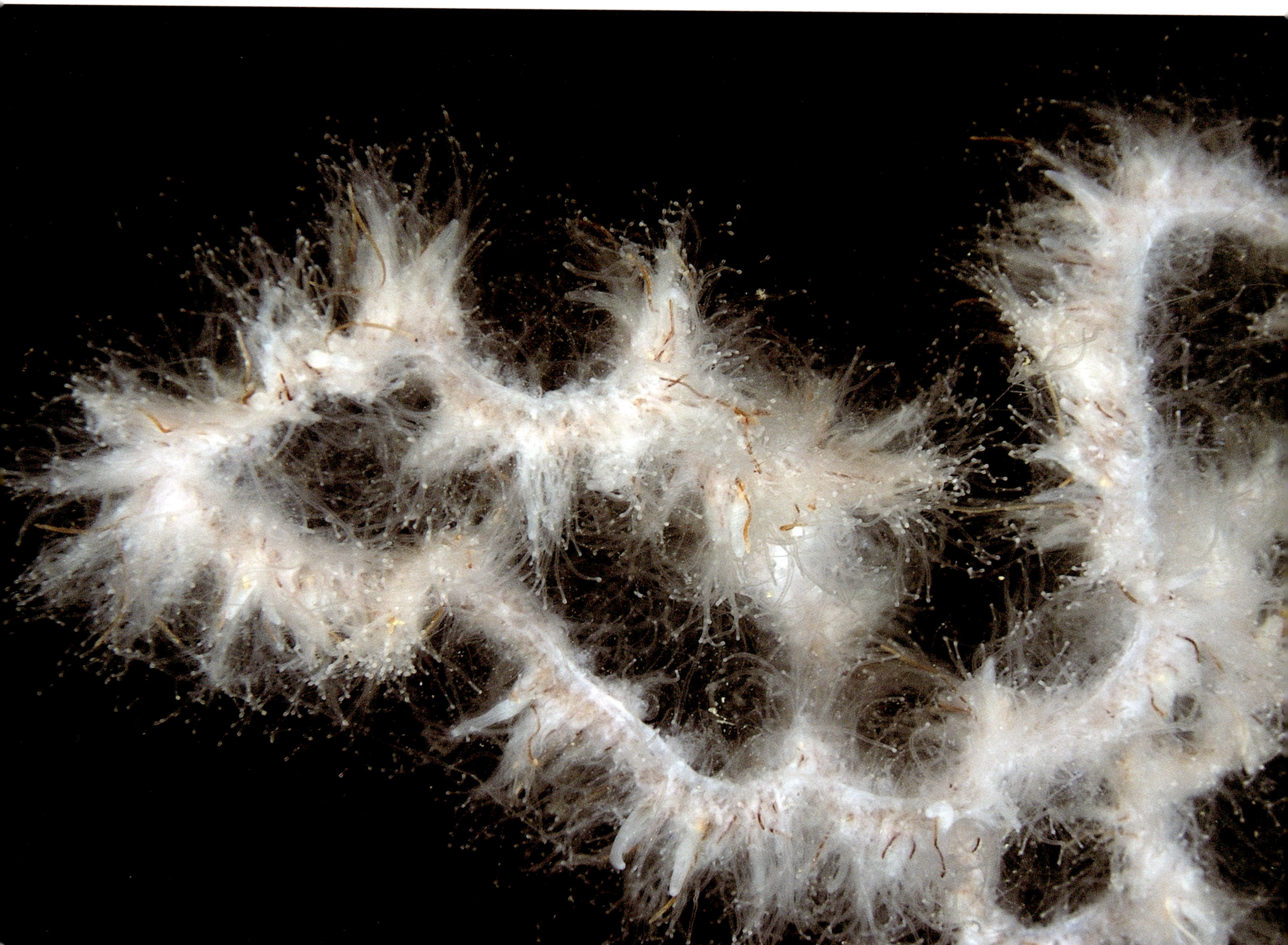

Pearl-chain Jellyfish, St. Finian's Bay, Co. Kerry

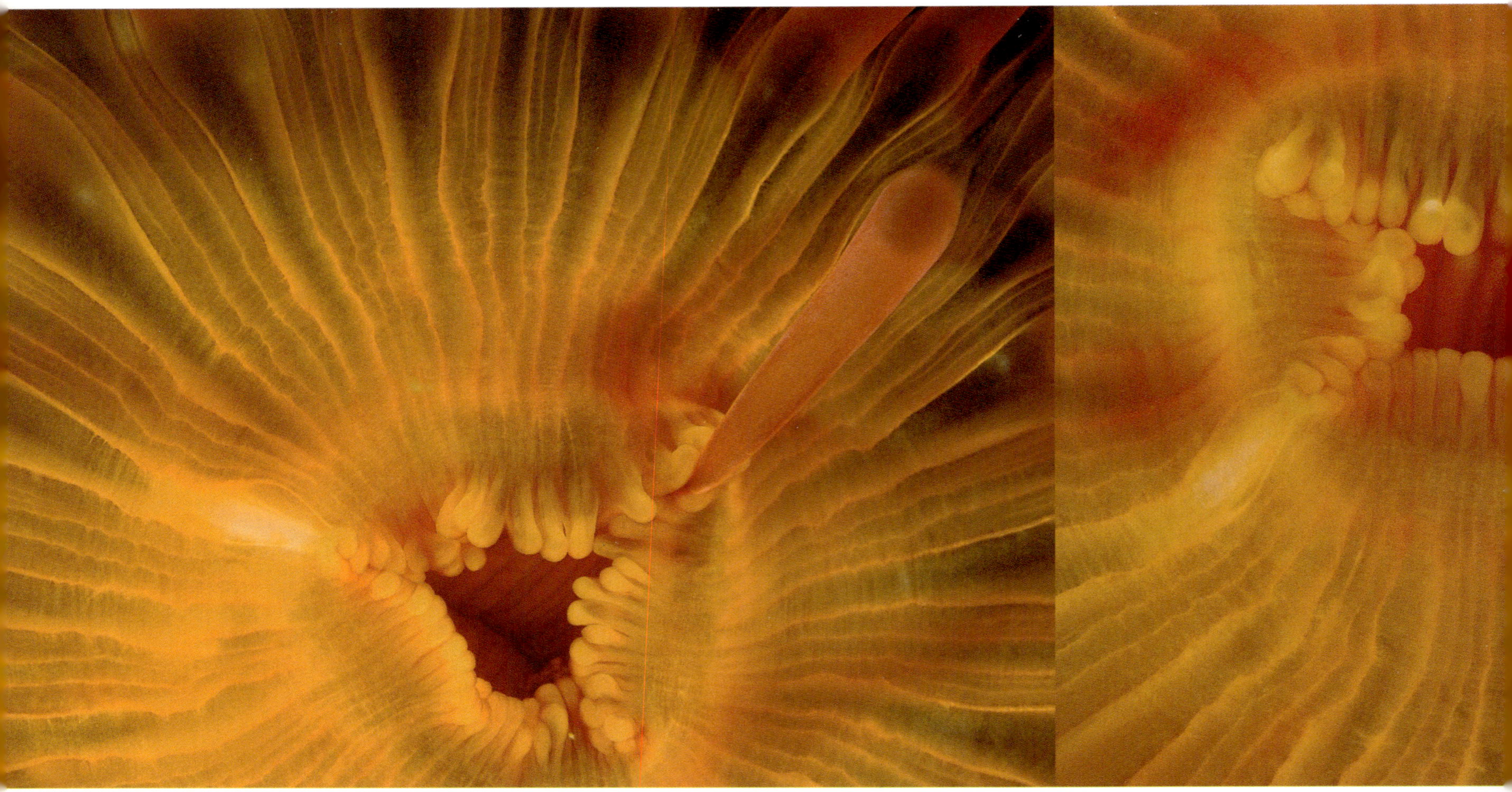

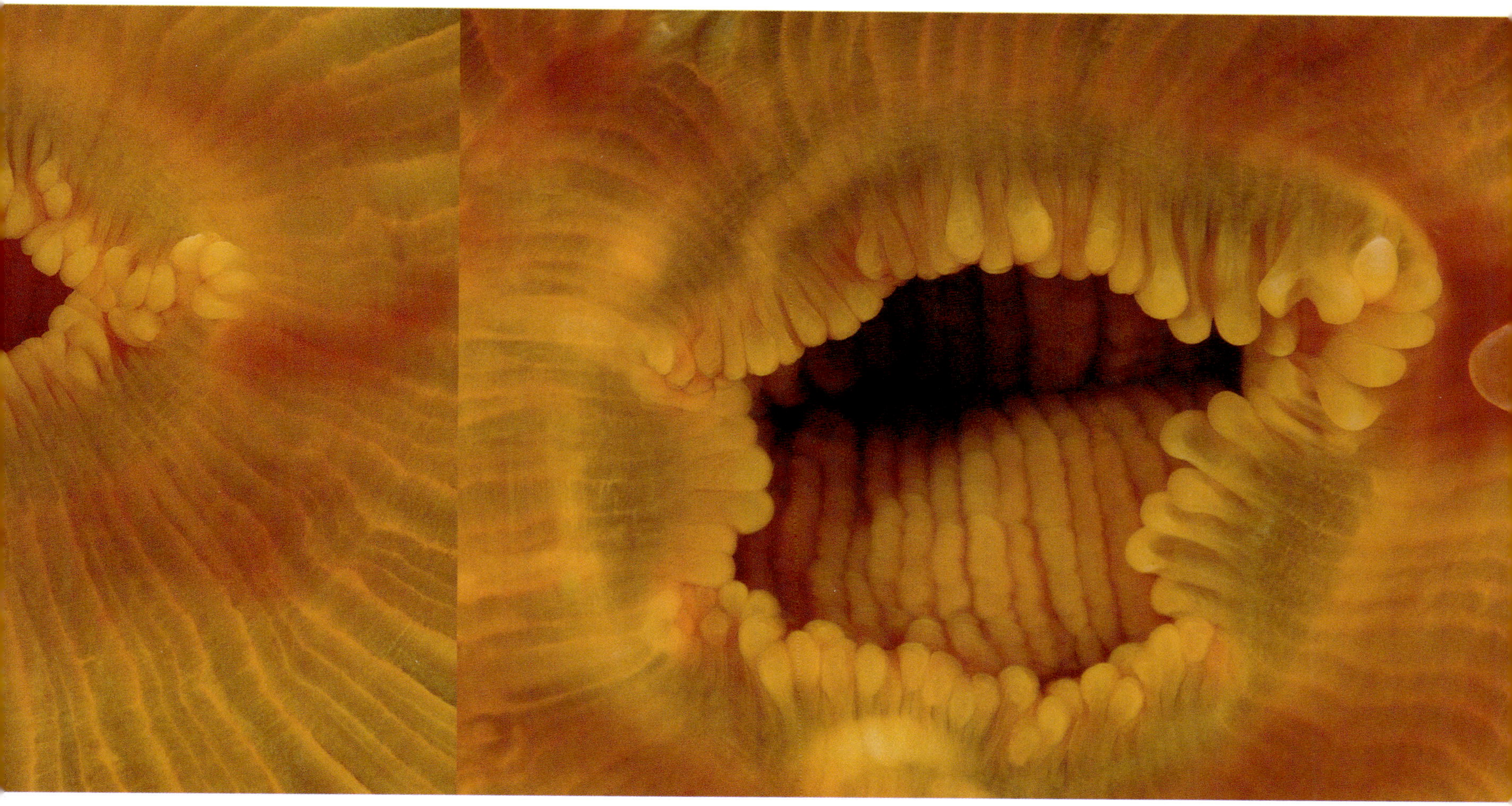

Anemone Feeding, Bull Rock, Cork

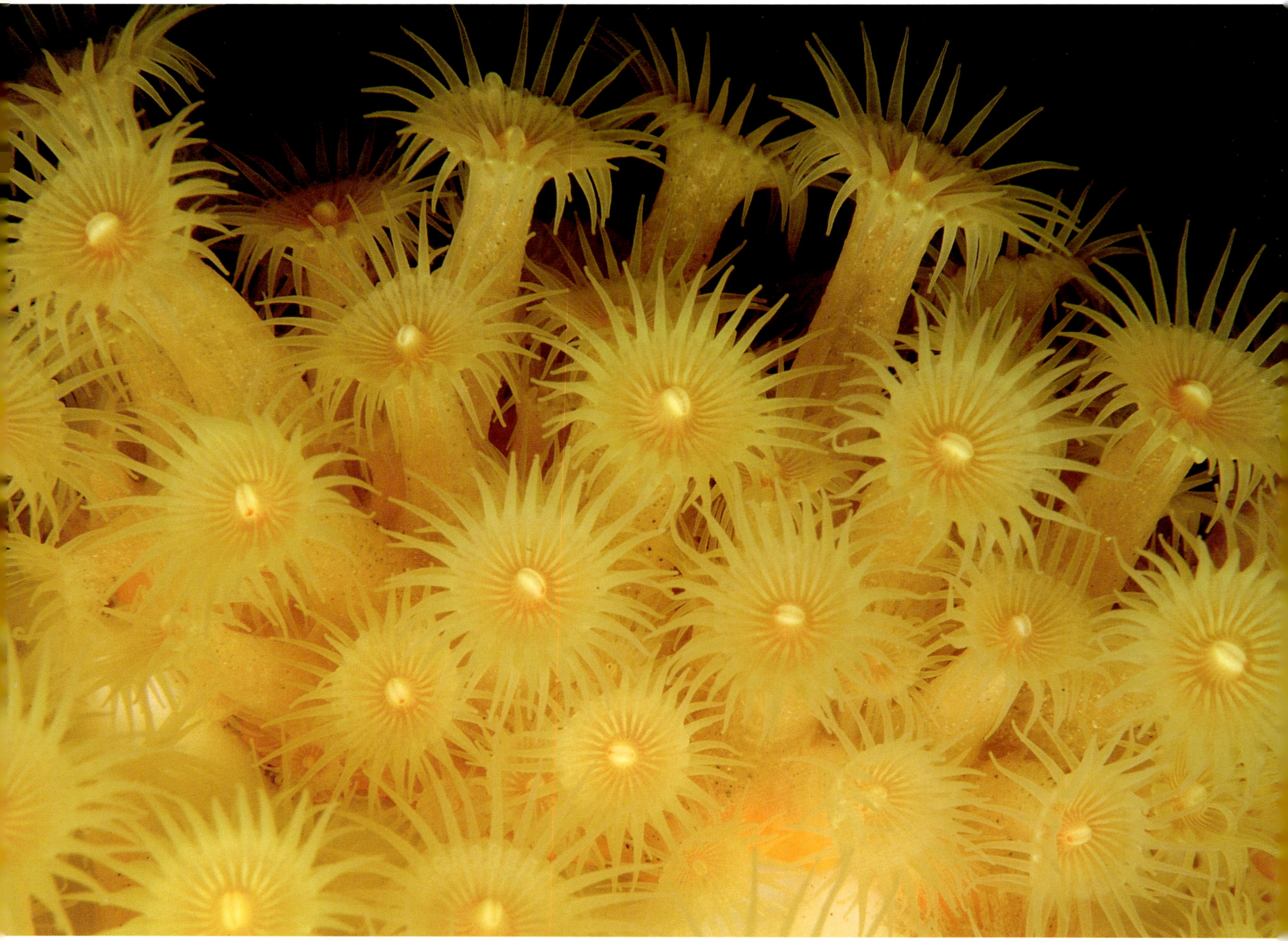

Yellow Zooanthids, Waterfall Bay, Tasmania

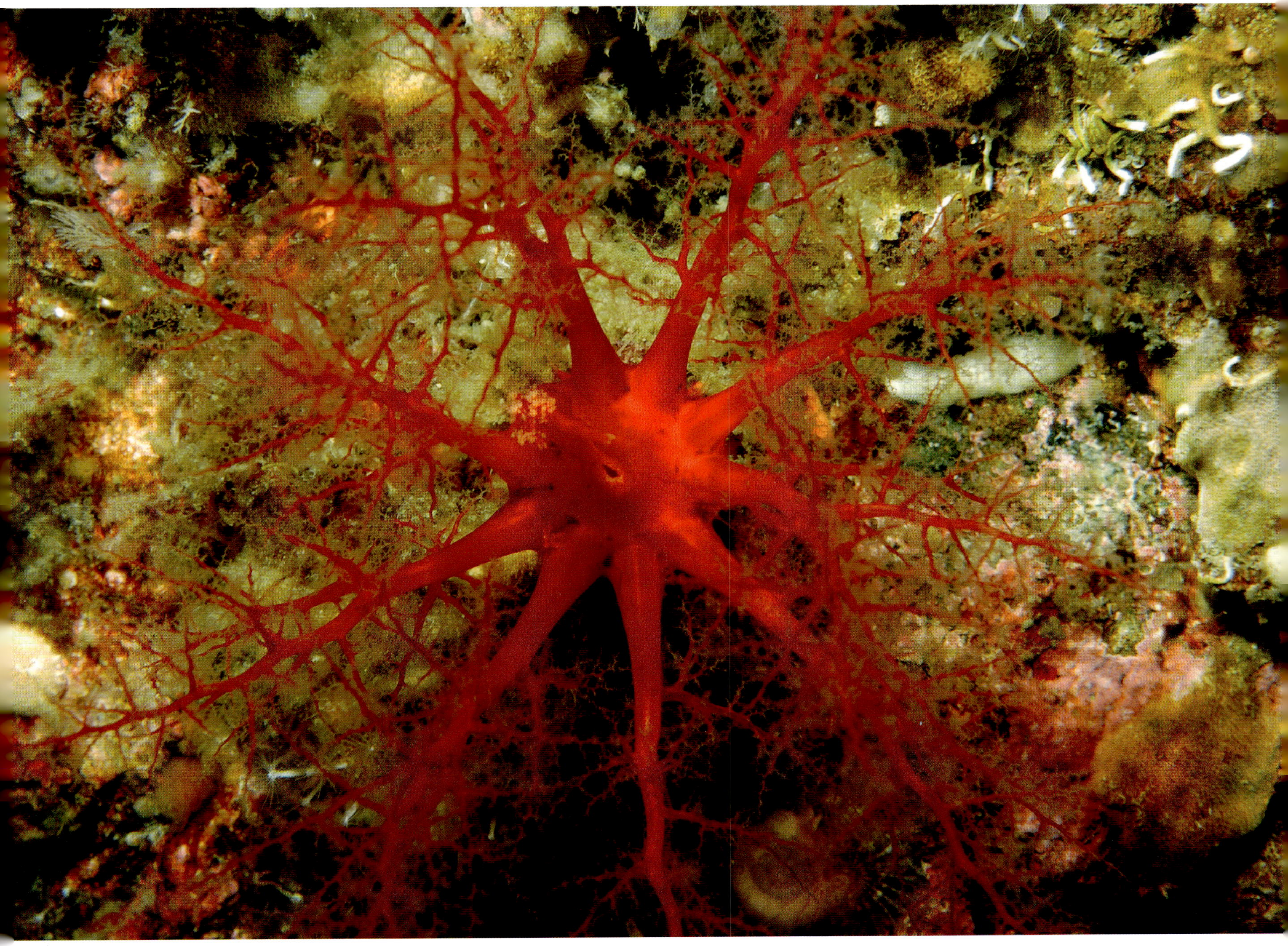

Sea Cucumber, Andersen Island, Vancouver

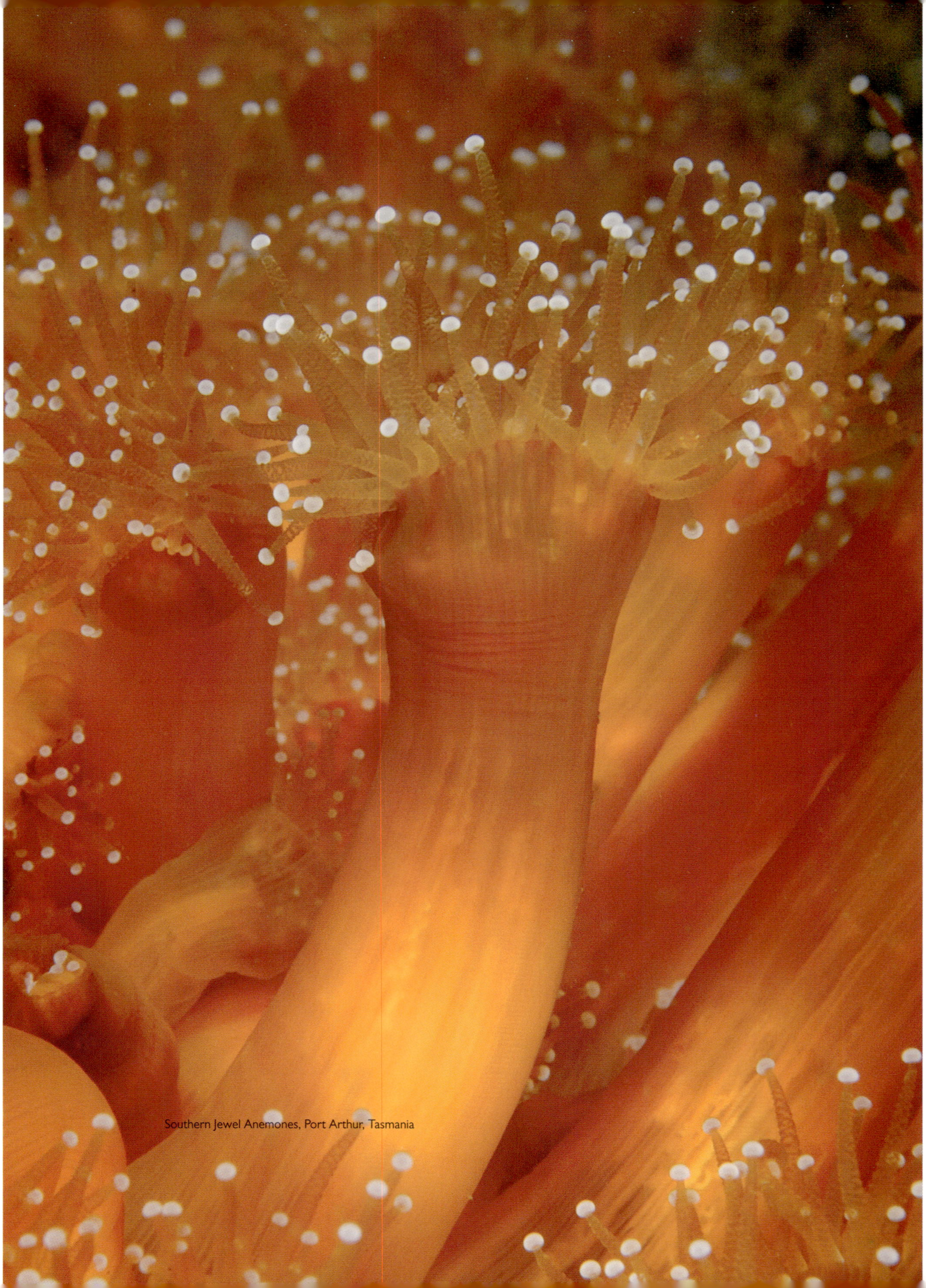

Southern Jewel Anemones, Port Arthur, Tasmania

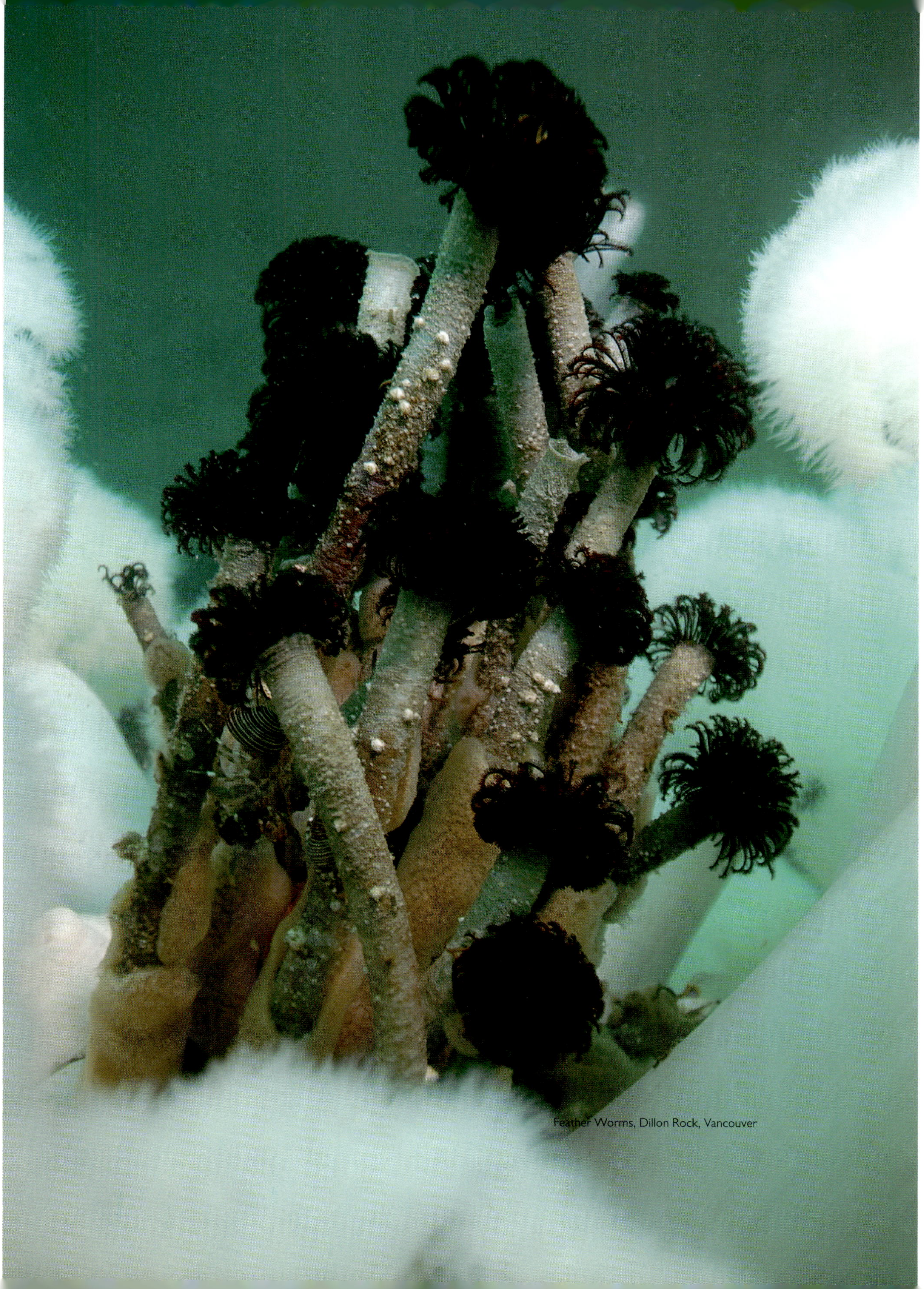

Feather Worms, Dillon Rock, Vancouver

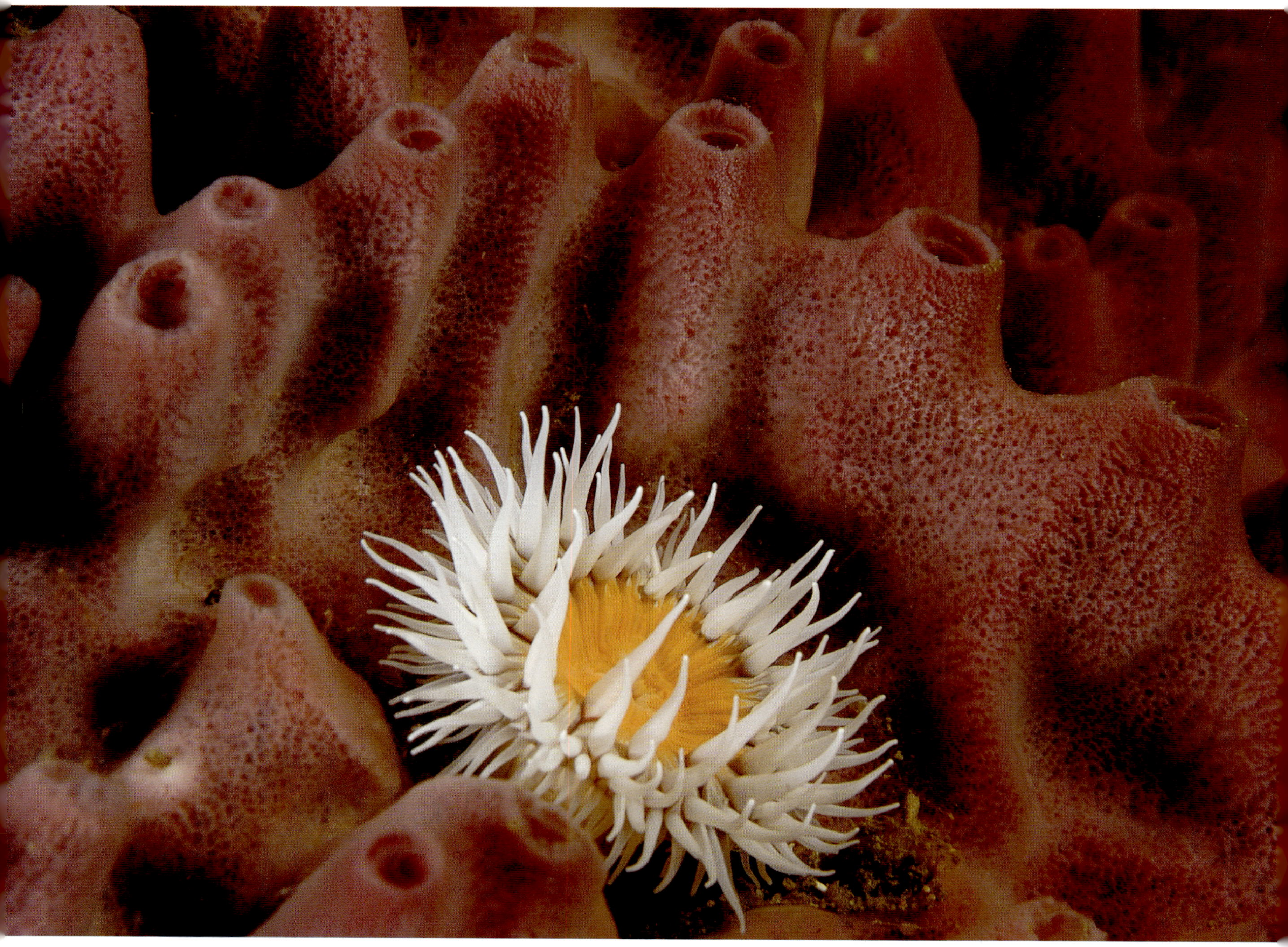

Anemone in Sponge, Bull Rock, Cork

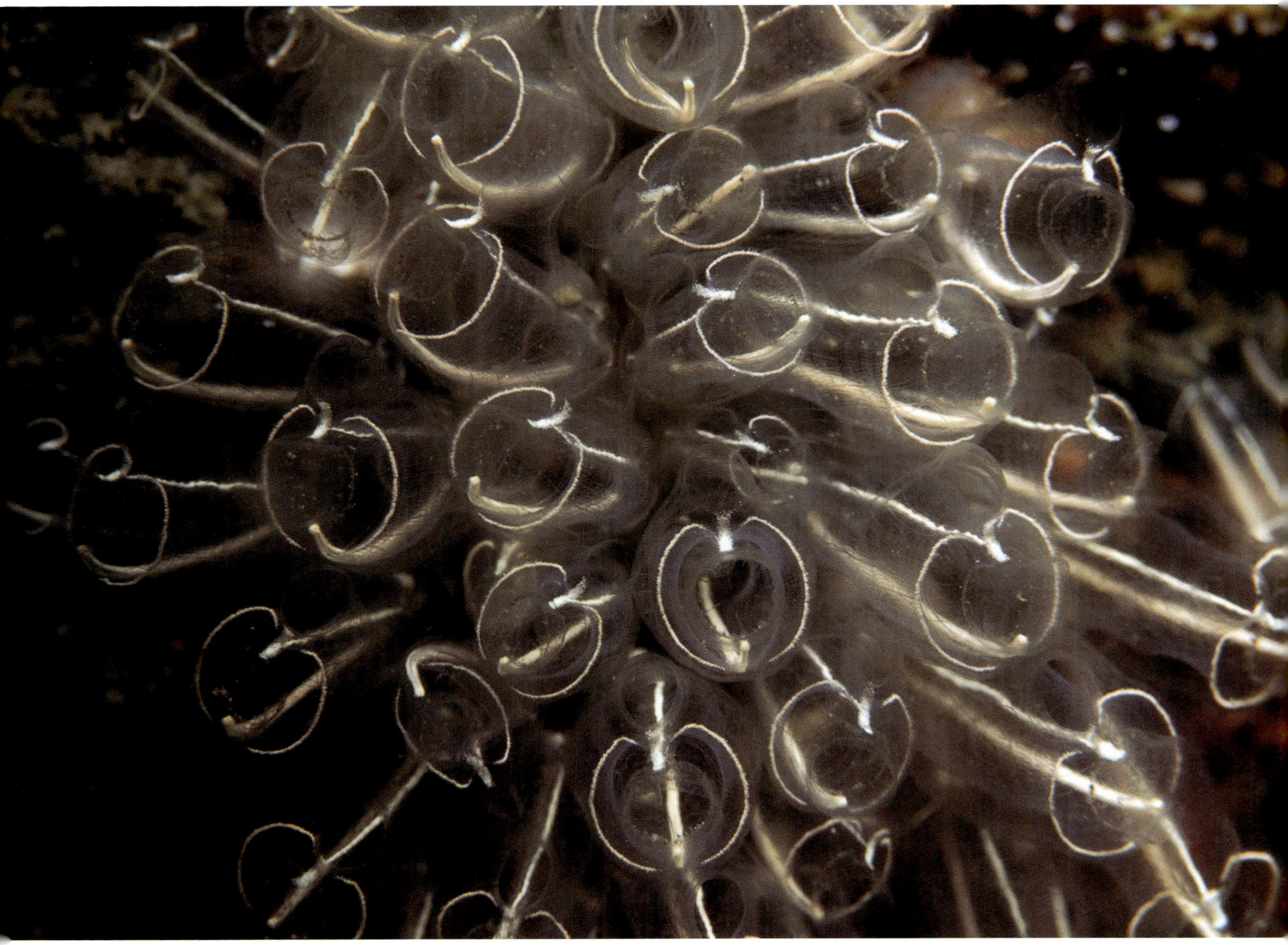

Light Bulb Sea Squirts, Lough Hyne, Cork

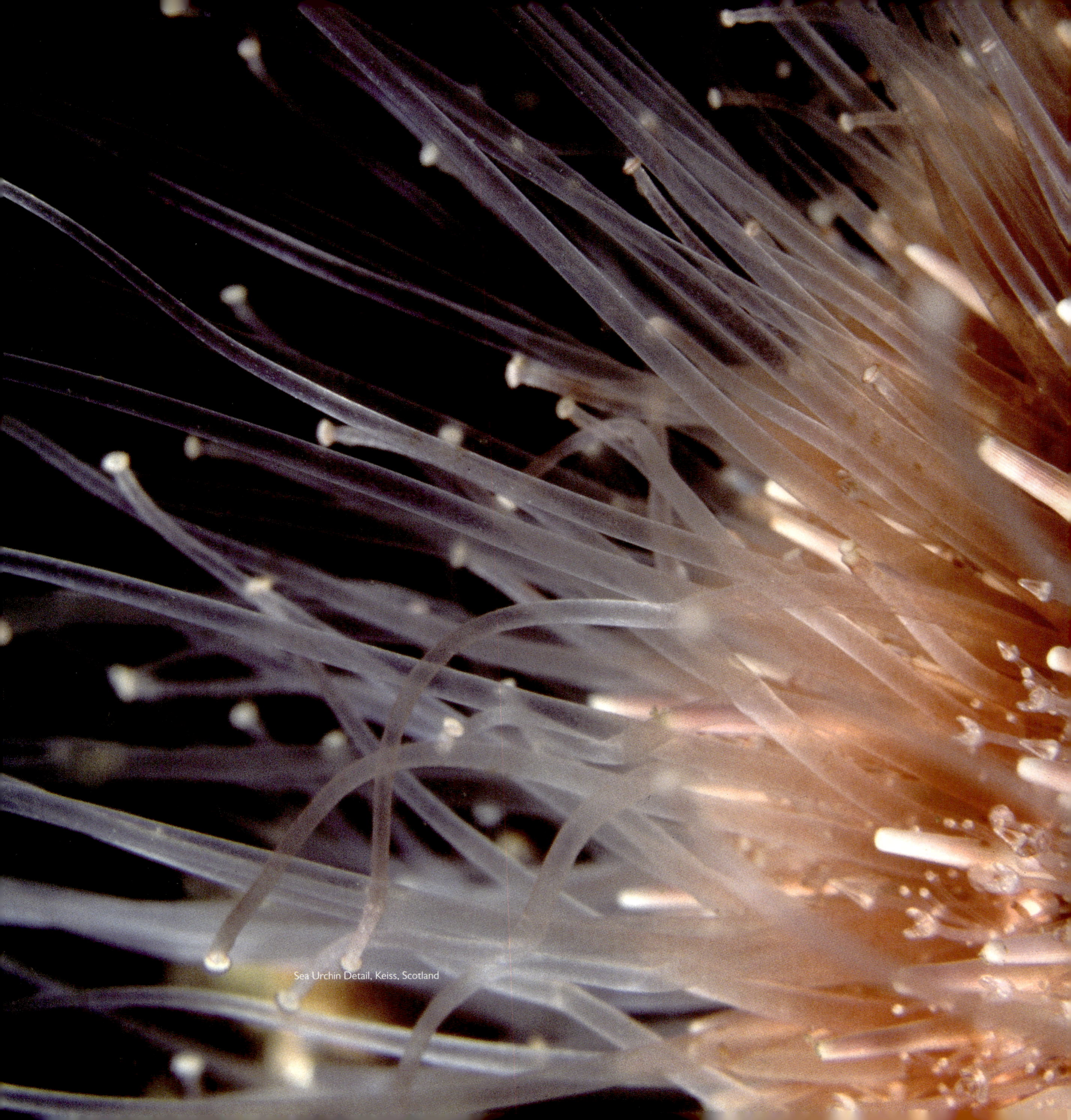

Sea Urchin Detail, Keiss, Scotland

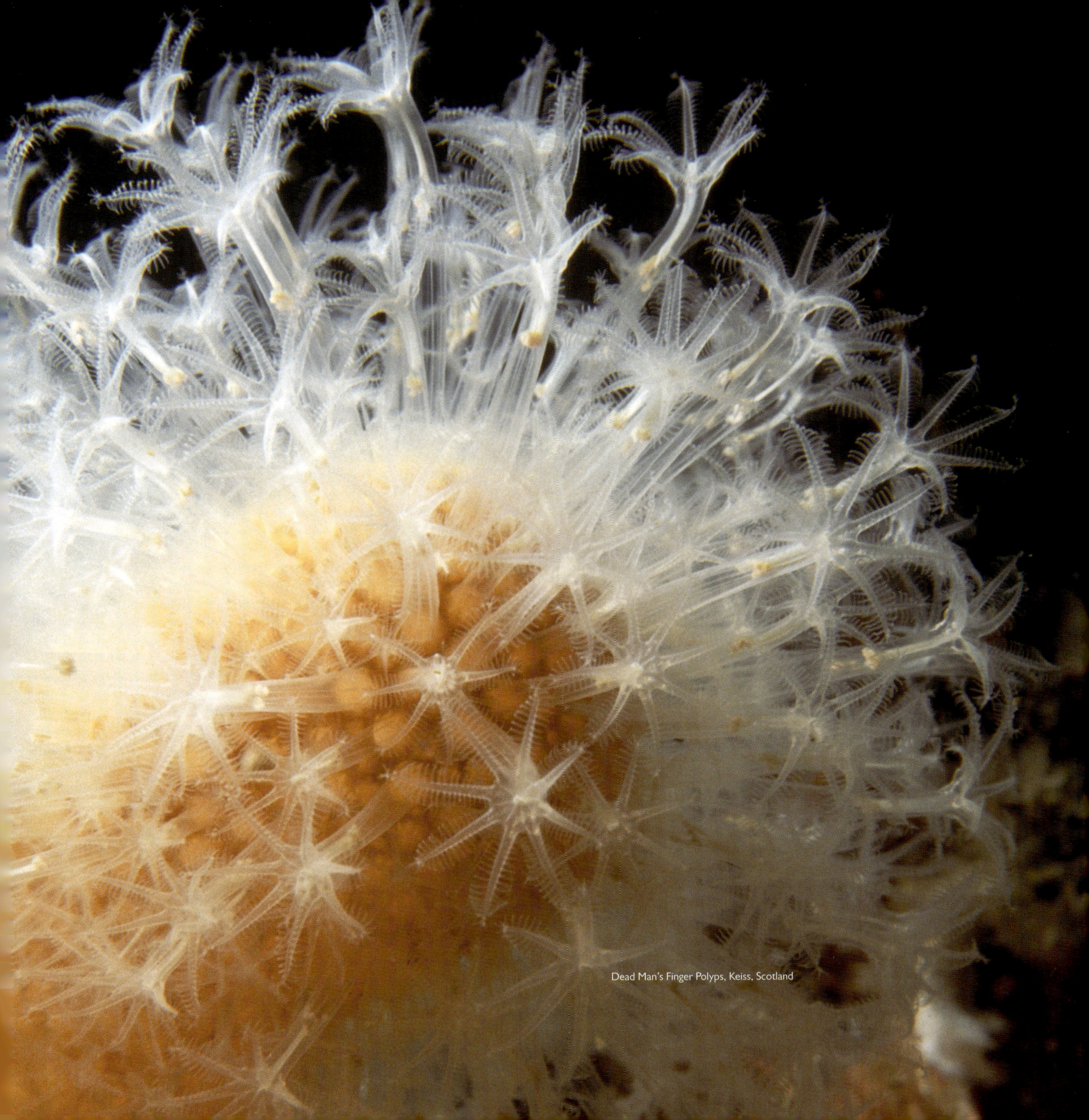

Dead Man's Finger Polyps, Keiss, Scotland

Valéry has written a line in 'Charmes', in which, with the foresight of true genius, he expresses the very essence of what I felt: 'Heureuse vos corps fondus, eaux planes et profondes!' (Happy your mingled atoms, O waters smooth and deep.) I was this 'corps fondu', the idealised but lively image of myself, and I moved in a landscape of dazzling reflections. The reverse side of the looking-glass was inhabited: fish, seaweed, now at eye-level, offered me the extension of a reality for which I was not prepared. I felt one by one, like a kind of arthritis, my land origins. The mind was less at home than the body. Freed of all anxieties about breathing, I moved around in the forbidden world…

Submarine poetry, which was only a dream to Valéry, is a favour that the sea accords right away to a diver of sensibility. He only needs a minimum of gifts: a certain skill in the water, a quality of awe, curiosity. A submarine sensibility will one day become one of the attributes of the cultivated man, like taste or an ear for music.

Philippe Diolé, *The Undersea Adventure*, 1953

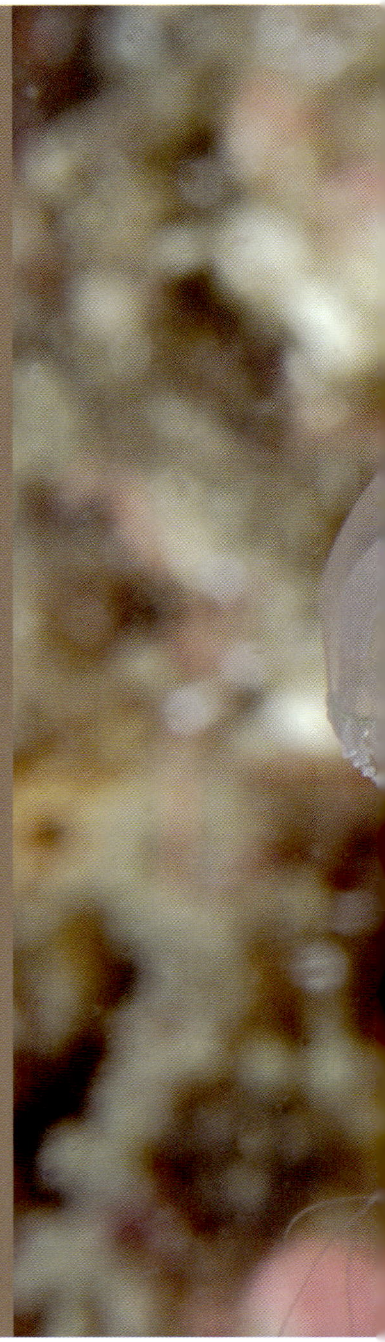

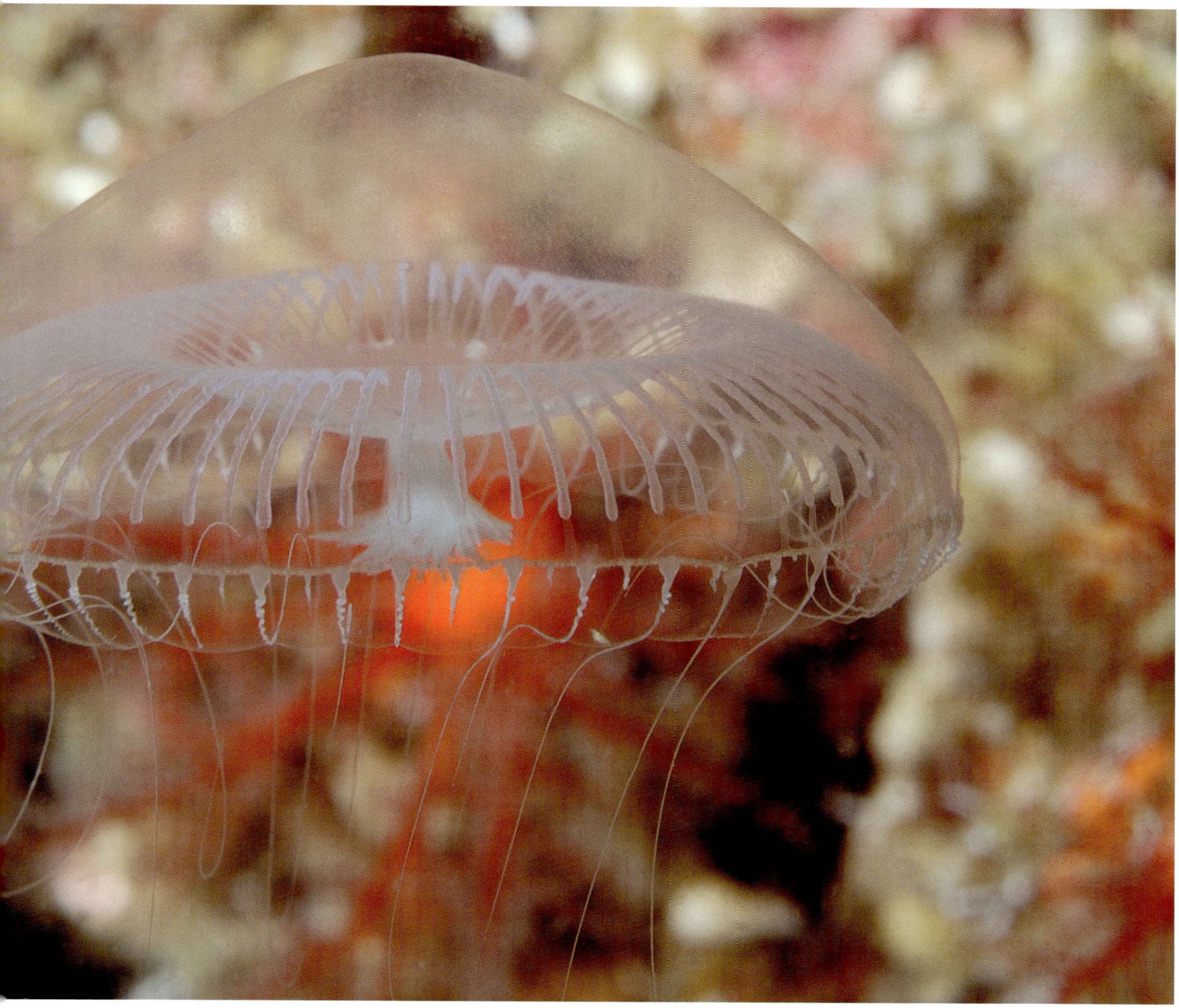

Jellyfish, Browning Pass, Vancouver

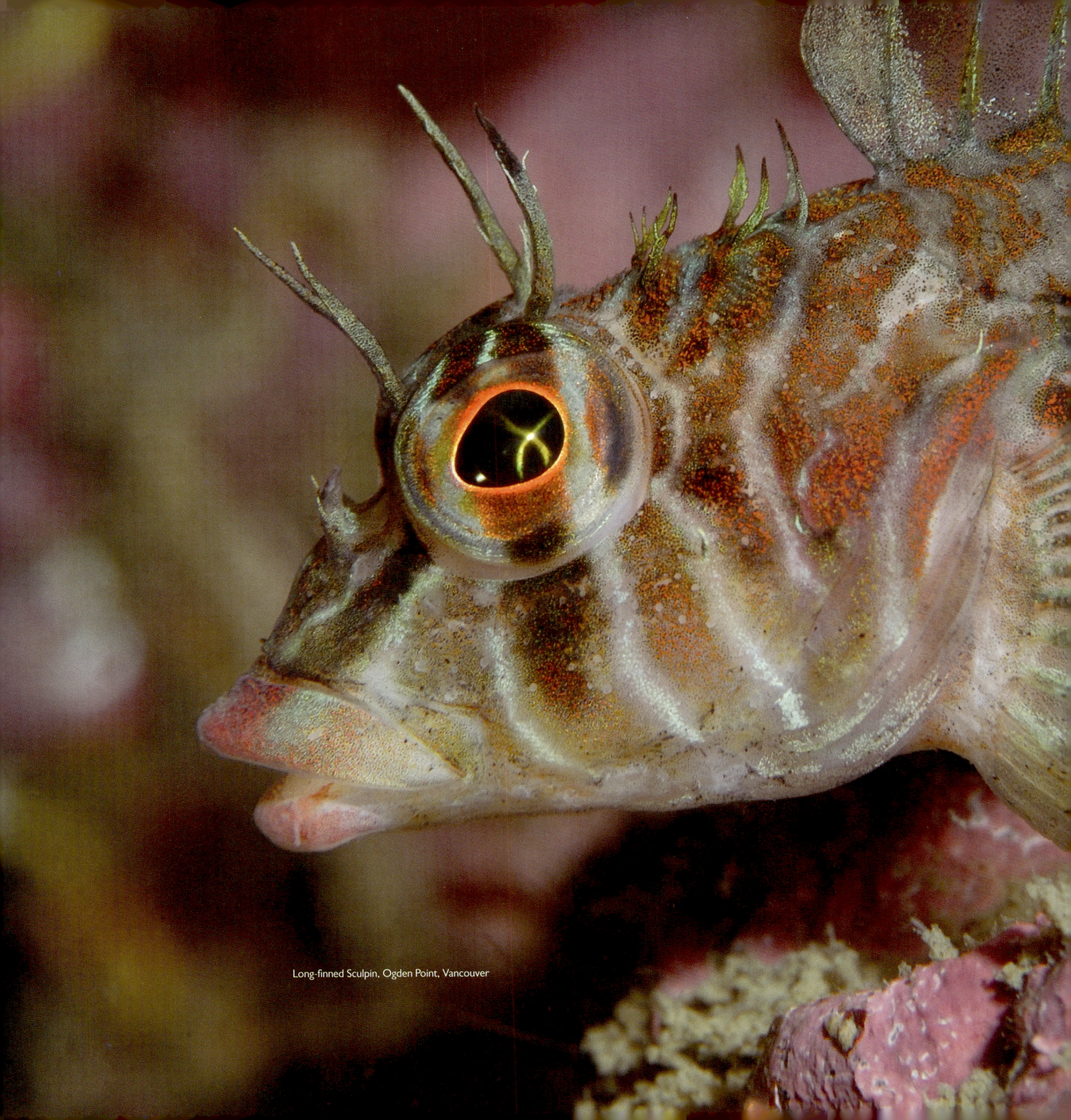

Long-finned Sculpin, Ogden Point, Vancouver

PHOTOGRAPHER'S NOTES

Long mysterious shadows filled the space between the ancient buildings, looming ghostly and unreal against the glow of the setting sun. Silence reigned. Above the crooked roofs and sagging chimneys was a fathomless green sky, and a strange sensation of standing at the bottom of the sea among the ruins of some sunken city came to me. I was seized with a sudden inspiration to make photographs of the world beneath the sea.

John Ernest Williamson, pioneer underwater photographer (1881–1966).

Capturing light in the sea is a peculiar passion. On land, you can watch and wait while the light changes, visualising the image that best captures the moment. The camera can be stabilised on a tripod to allow careful, considered compositions. Several rolls of film, or memory cards can be exposed with different lenses and evaluated later to choose the image that best expresses the picture seen in the mind's eye. Diving with a camera is a completely different endeavour and an equation of many variables.

Diving skills and safety come first. Self-reliance, ocean knowledge and a sixth sense of awareness on the dive are essential before we add photographic skills and artistic vision. Unlike working on land, you can carry only essential equipment – life support for the environment, instruments to monitor the dive and a limited supply of photographic tools. People are surprised to hear that it is generally land equipment that we bring underwater, in specialised camera housings. The best designs allow use of all essential camera controls, yet are ergonomic and intuitive to use. A commitment to a certain type of image must be made before the dive – either wide angle views encompassing large areas of reef or close-up details of the animal life of the sea. Once this decision is made, the camera housing is sealed and it must remain so until back on dry land, so lenses, film or memory cards cannot be changed underwater.

The single most challenging element in emerald seas, however, is water itself as life-giving plankton soaks and scatters light. At depth, everything is bathed in a green glow, shapes visible only for a few metres. Even the clearest water filters out colour: first reds, then orange and yellow; eventually only blue, green and grey remain. To appreciate the underwater world, we must bring light in the form of dive lights and underwater flash units. Good photography uses these sparingly, to retain authenticity – existing light with a 'kiss of flash'.

Over the years, I have used many different equipment configurations, starting with amphibious Nikonos cameras in the 1980s and most recently, professional Nikon Digital cameras. While the early camera systems were compact and easy to work with, the learning curve was a steep one. Moving to housed cameras in the 1990s was a major step forward in having reflex viewing to allow accurate composition. Technological progress in light measurement, autofocus and motorised film advance also made life a lot easier. But digital technology has

It takes a lot of imagination to be a good photographer. You need less imagination to be a painter, because you can invent things. But in photography everything is so ordinary; it takes a lot of looking before you learn to see the ordinary.

David Bailey, *Face*

revolutionised photography in the early years of the 21st century. Not only do we have instant feedback on lighting and exposure but large capacity storage that allows hundreds of images to be shot on a single dive. The limitation of 36 film exposures underwater seems a distant memory already. Working digitally is also a lot more enjoyable, seeing results immediately and being able to review the dive's shoot while decompressing. And while learning to use the computer as a darkroom has been challenging, it has been hugely rewarding.

The drive to capture and share images of the sea is compelling. On a crowded city street, I have often wondered how many of the faces passing by know about perfect pink sea-anemones, the dance of a dolphin or time-capsule shipwrecks at the bottom of the sea. It gives me the fuel to persevere in creating intimate portraits of the sea and its magnificent inhabitants.

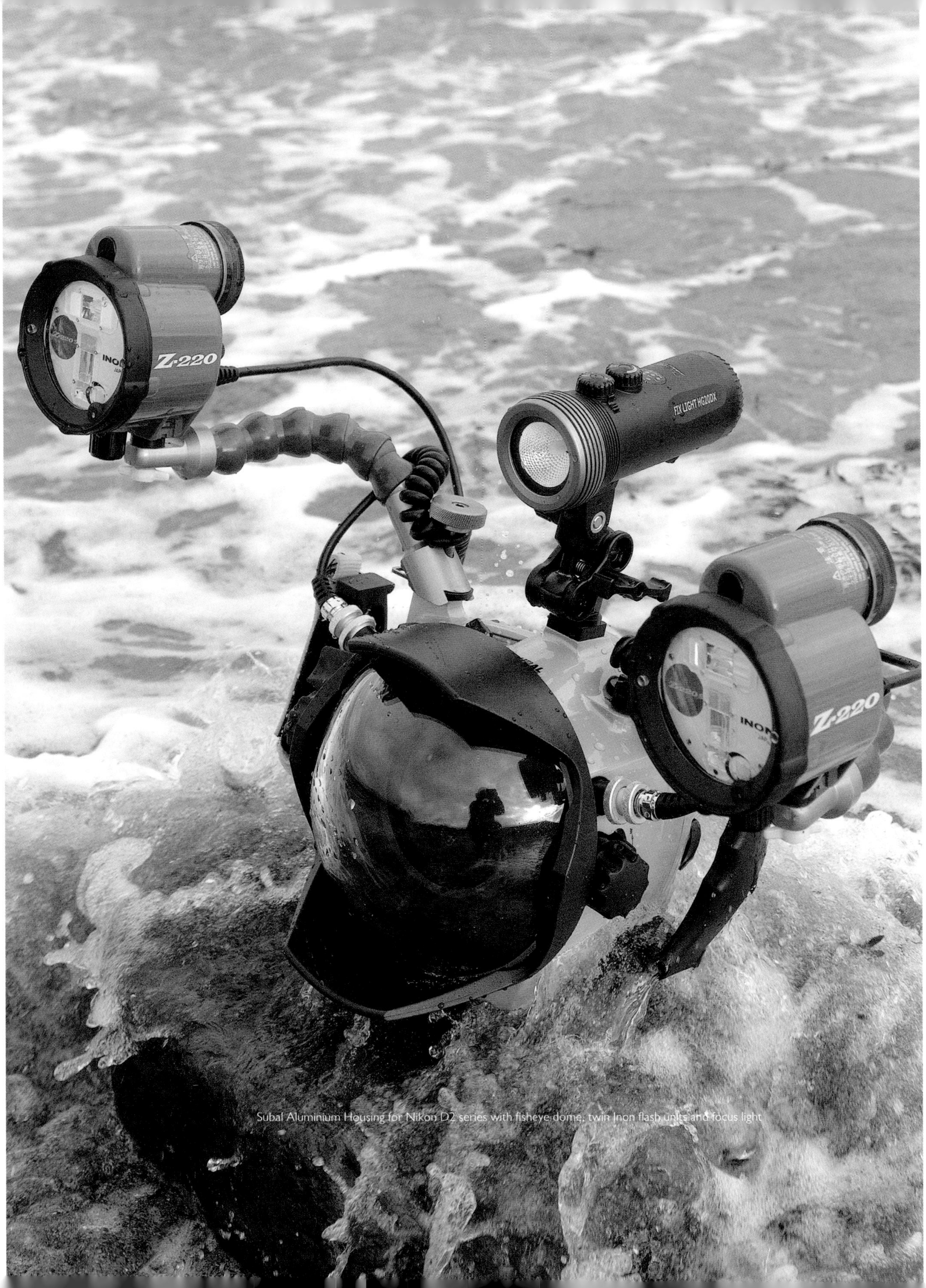

Subal Aluminium Housing for Nikon D2 series with fisheye dome, twin Inon flash units and focus light

Cousteau had, even more enthusiastically than before, taken up his undersea activities again. He had rigged up an old movie camera, picked up in a junk shop in Marseilles, a 35mm Kinamo, with a first-class lens. Movie films having disappeared from the market, he had made shift with some Leica spools, which were still freely obtainable in the shops. For this heaven-sent camera Véche made a watertight box, which incorporated the fruits of our by now long experience. Provided with a glass bull's-eye in front of the lens and stuffing boxes giving access to the controls, it was expected to withstand the pressure of sixty feet of water, the maximum depth of our skin dives. Like our weapons, it was fitted with a handle and grips, its trigger, instead of shooting arrows, releasing the film.

Hunters of submarine pictures today, with all kinds of highly developed cameras available to them, perhaps have little idea of the difficulties, the disappointments, the horrible waterlogging of our apparatus which accompanied our first pioneer experiments.

Captain Philippe Tailliez, *To Hidden Depths*, 1954

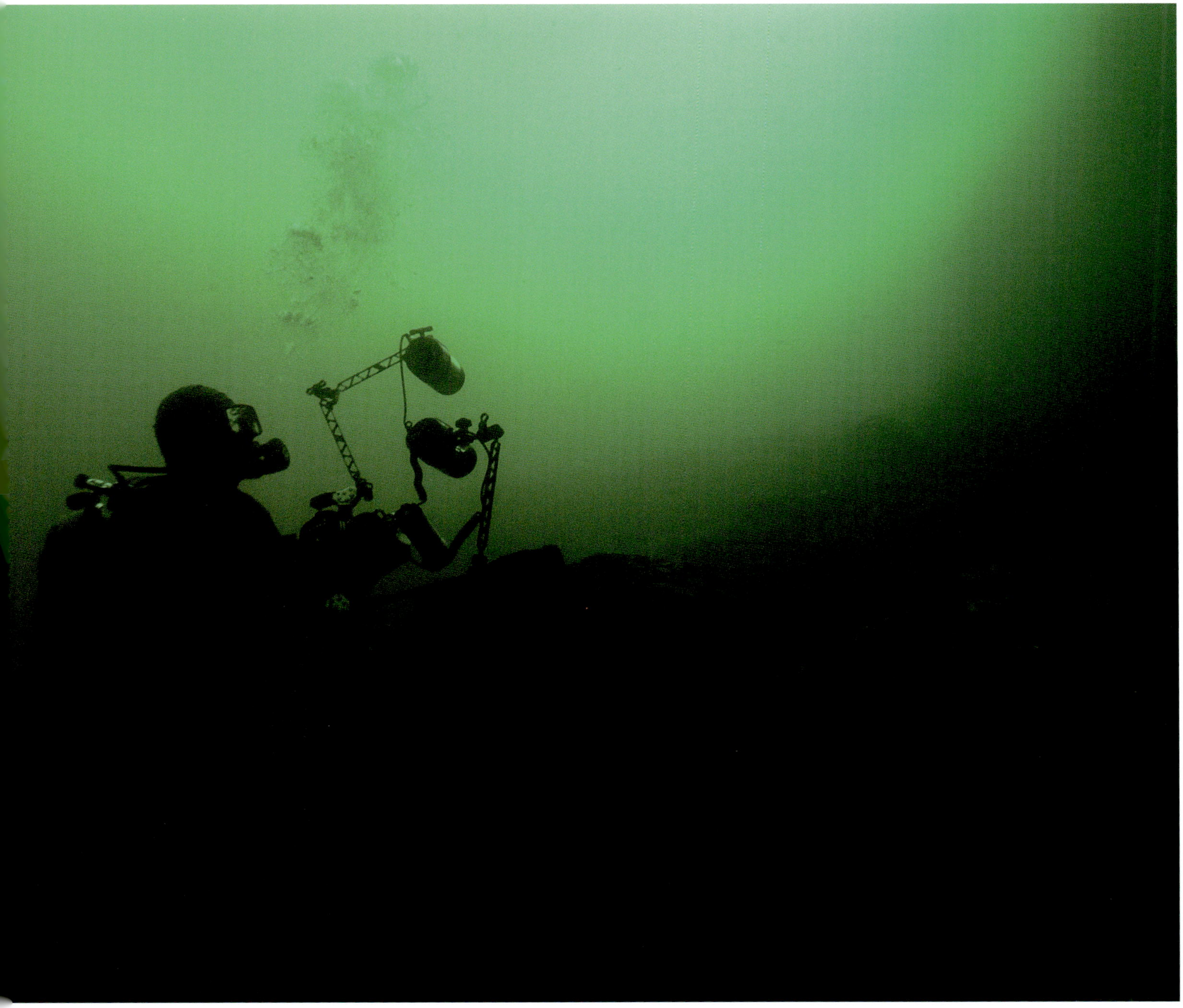

Emerald Sea, Tahsis, Vancouver

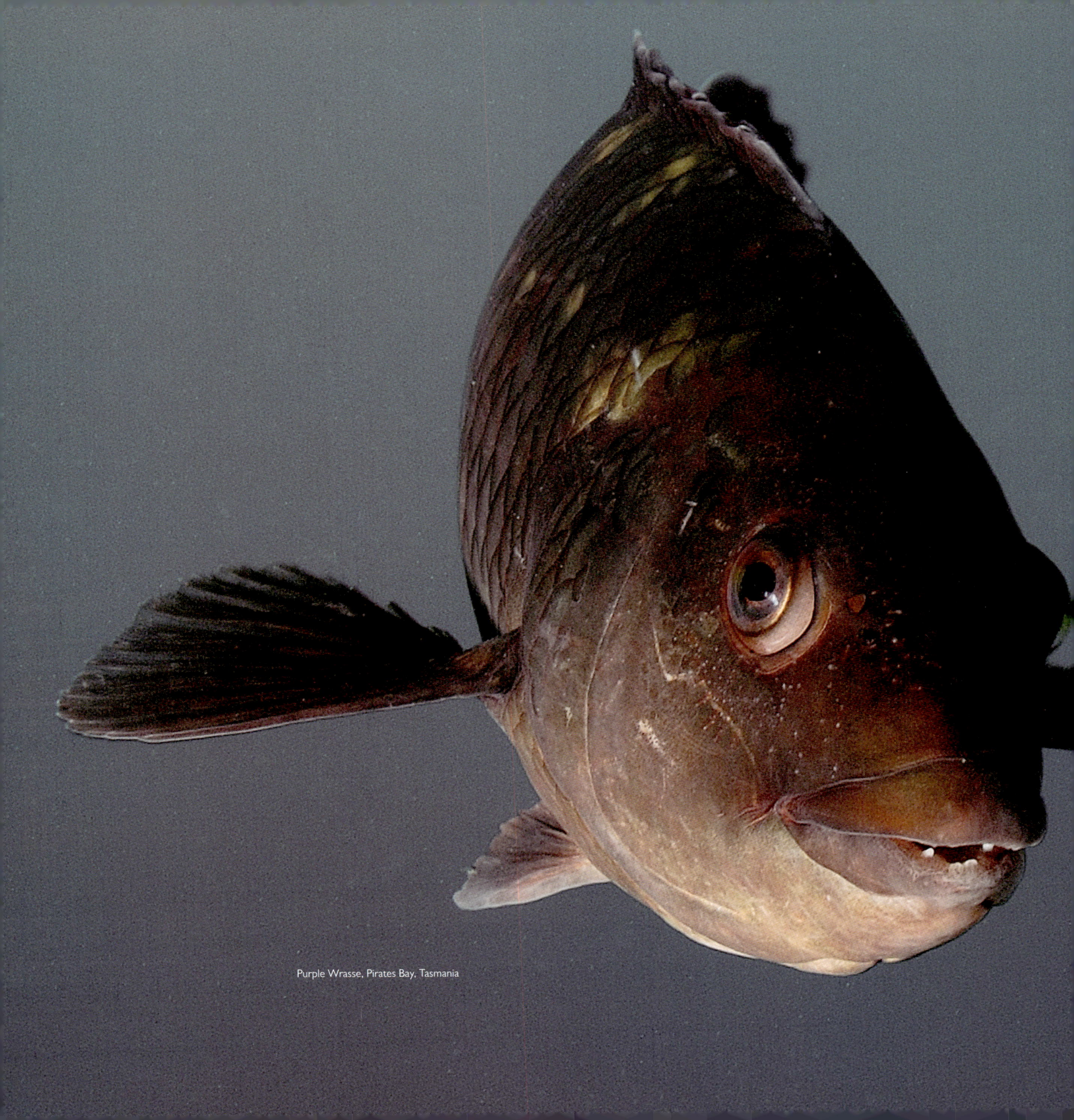

Purple Wrasse, Pirates Bay, Tasmania

ACKNOWLEDGEMENTS

Those closest to us are the most intimately involved in bringing a project such as this to fruition. My heartfelt thanks to my wife Caoilfhionn for the space and encouragement to complete the book and to Aislinn and Éadaoin – my biggest fans. To my parents Marion and Alphy, my sister Annmarie, brother Brian, and the McDonnell family who witnessed the growth in camera bodies and dive gear over the years and were always enthusiastic about the images.

While only one finger presses the shutter on the camera, many hands support an underwater photographer. Diving is at once an entirely personal and individual experience, and at the same time a team effort to carry out safely. To my countless diving companions over the years, thank you for your patience. Special thanks to my buddies, Caoilfhionn, Shirley, John, Tony, Pat and Aidan – your constant support is deeply appreciated.

Putting to sea safely is an integral part of diving and sincere thanks to all those who have skippered boats and waited anxiously for divers to return to the surface. Diving is also a constant learning practice, building gradually on a solid foundation. I am grateful to my early instructors and friends in Dublin University Sub-Aqua Club and subsequent colleagues in Cork Sub-Aqua Club. For later technical diving and rebreather training, I am indebted to Tim Doyle and Lee Crabtree. I have no doubt that their attitudes and standards have prevented me from becoming complacent and compromising my safety.

In diving new places, I am grateful to those who took the effort to brief me well and ensure that the diving was enjoyable, safe and productive: in Vancouver, Mike Lever, Berkley White, Phil Sammet and the crew of the *Nautilus Explorer*; in South Africa, Michael Scholl and the crew of *White Shark*; in Tasmania, Mick Baron, David Perkins and Ian 'Blakey' Blakeway of Eaglehawk Dive Centre. Closer to home in Ireland, thanks to Nick Pfeiffer on the Aran Islands and Willie and Sarah Kennedy in St. Finian's Bay, Co. Kerry. And in fond memory of Nic Gotto who always ensured that I had the first few minutes of clear visibility on U-260 before a boat-load of divers descended on it.

Underwater photography is an unusual passion, to say the least and sharing knowledge and experience is fundamental to learning. I am particularly grateful to Nigel Motyer for his selfless sharing of hard-won experience and technical expertise. I am also grateful to Steve Warren of Ocean Optics, Berkley White of Backscatter and and to the many underwater photographers that I have met and shared ideas with over the years. Finally, to my friend Maria O'Donovan who opened the door to the mysterious world of publishing and to Niall McAuliffe for his brainwave which led to me Cork University Press. Sincere thanks to Tom Dunne and all of the team at CUP in seeing the book through to production, and especially to John Foley of Bite! Design whose creativity has been an inspiration.

JC, May 2006.

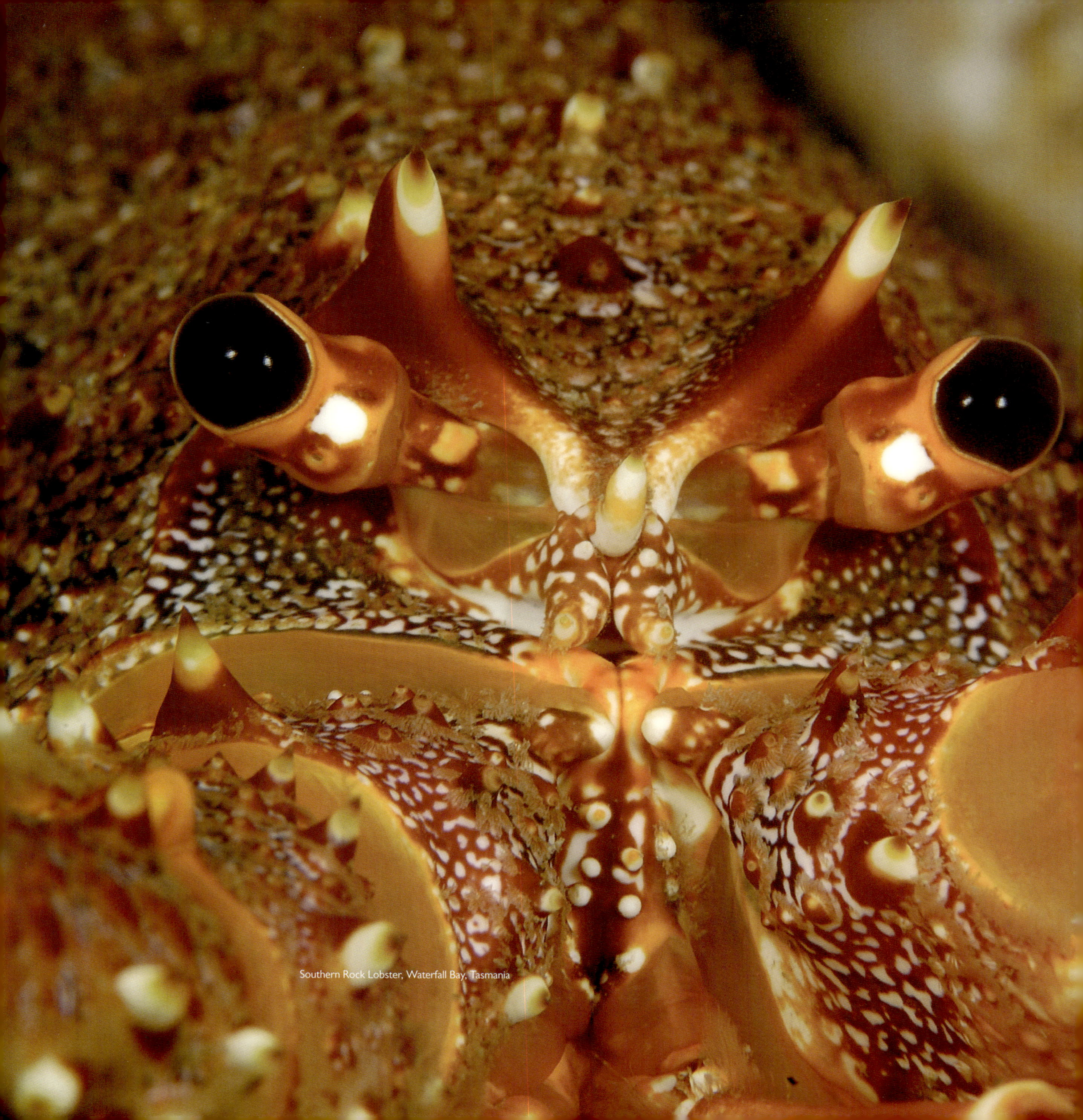

Southern Rock Lobster, Waterfall Bay, Tasmania

FURTHER READING

Beebe, William, Beneath Tropic Seas: A Record of Diving among the Coral Reefs of Haiti, (New York: Putnams, 1928).

Benchley, Peter, Shark Trouble, (New York: Random House, 2002).

Byatt, Andrew et al, The Blue Planet, (London: BBC Books, 2001).

Carson, Rachel L., The Sea Around Us, (New York: Oxford University Press, 1951).

Carwardine, Mark and Ken Watterson, The Shark Watcher's Handbook, (London: BBC Worldwide Ltd, 2002).

Cousteau, J. Y. and Dumas, Frederic, The Silent World, (New York: Harper and Rowe, 1953).

Cousteau, Jacques with James Dugan, The Living Sea, (New York: Harper and Rowe, 1963).

Cousteau, Jacques, The Ocean World, (New York: Harry N. Abrams, 1979).

Diole, Philippe, The Undersea Adventure, translated by Alan Ross. First published under the title L'Adventure Sous-Marine, by Albin Michel, Paris, 1951. First English translation 1953, published in Great Britain by Sidgwick and Jackson Ltd, London. Also published in the USA in 1953 by Julian Messnr, Inc, New York.

Doubilet, David, Light in the Sea, (London: Swan Hill Press, 1989).

Doubilet, David, Water Light Time, (London: Phaidon Press, 1999).

Dugan, James, Man Explores the Sea, (London: Hamish Hamilton, 1956).

Ecott, Tim, Neutral Buoyancy: adventures in a liquid world, (London: Michael Joseph (Penguin), 2001).

Edge, Martin, The Underwater Photographer: Digital and Traditional Techniques, (Oxford: Focal Press, 2005).

Ellis, Richard and McCosker, John E., Great White Shark, (Stanford University Press, 1991).

Hass, Hans, Men Beneath the Sea, (New York: Saint Martin's Press, 1975).

Kay, Paul, Underwater Photography, (London: Guild of Master Craftsman Publications, 2003).

Jackson, Jack, Diving with Sharks and Other Adventure Dives, (London: New Holland Publishers, 2000).

Naylor, Paul, Great British Marine Animals, (Sound Diving Publications, 2005).

Newbert, Christopher, Within a Rainbowed Sea, (Hawaii: Beyond Worlds Publishing, 1984).

Newbert, Christopher, and Birgitte Williams, In a Sea of Dreams, (Texas: Fourth Day Publishing, 1994)

Seaborn, Charles, Underwater Photography, (New York: Amphoto, 1988).

Skerry, Brian and Howard Hall, Successful Underwater Photography, (New York: Amphoto Books, 2002).

Tackett, Denise Nielsen and Larry Tackett, The Essential Underwater Photography Manual: A Guide to Creative Techniques and Key Equipment, (Hove, UK: Rotovision SA, 2005).

Tailliez, Philippe, To Hidden Depths, (London, William Kimber, 1954).

Verne, Jules, 20,000 Leagues Under the Sea, (Paris: Hetzel, 1870).

Webster, Mark, The Art and Technique of Underwater Photography, (London: Fountain Press, 1998).

Wu, Norbert and Ken McAlpine, Diving the World, (New York: Hugh Lauter Levin Associates, 2003).

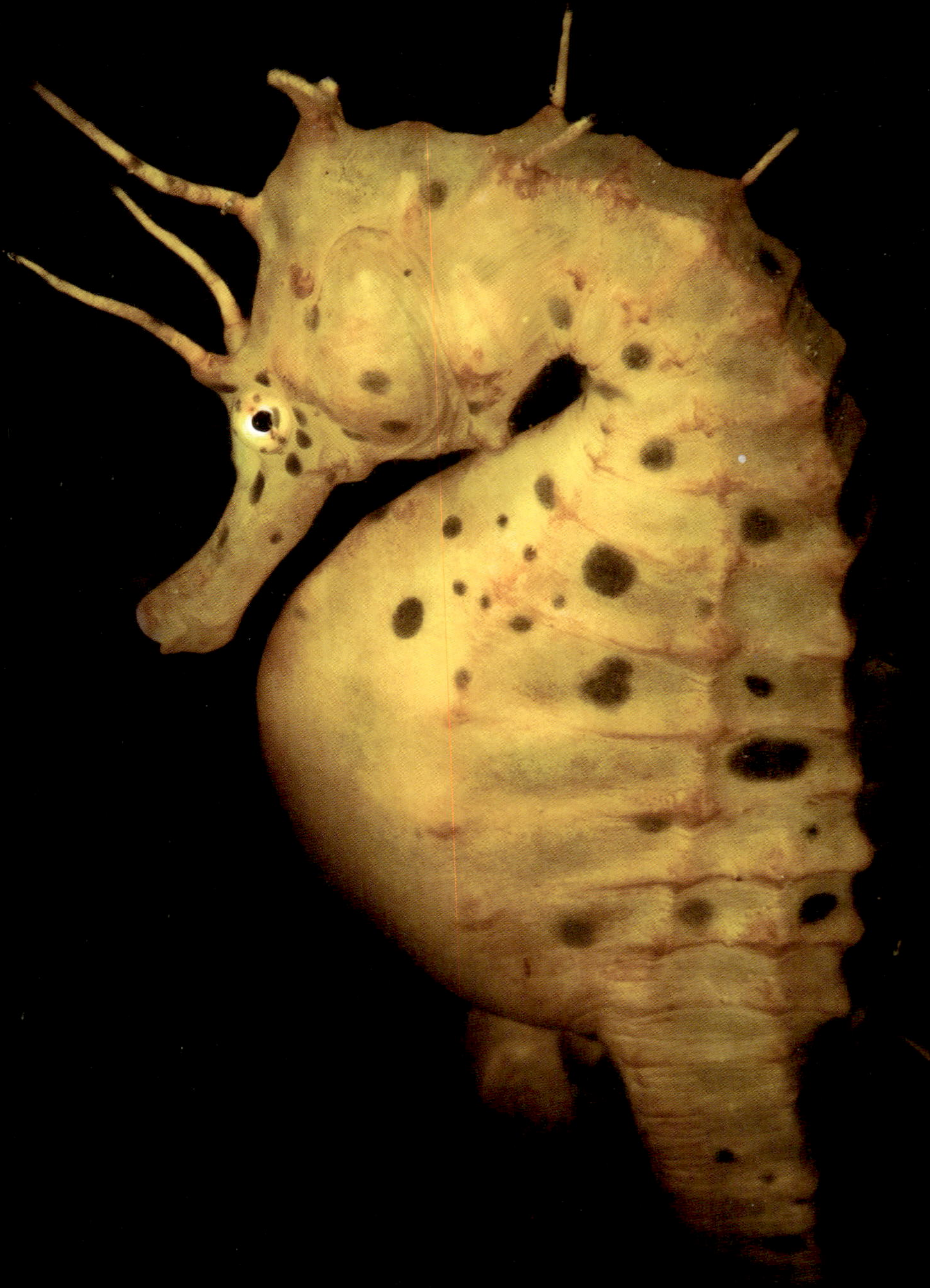

Pot-bellied Seahorse, Pirates Bay, Tasmania

Endpapers: Star Squirt Abstract

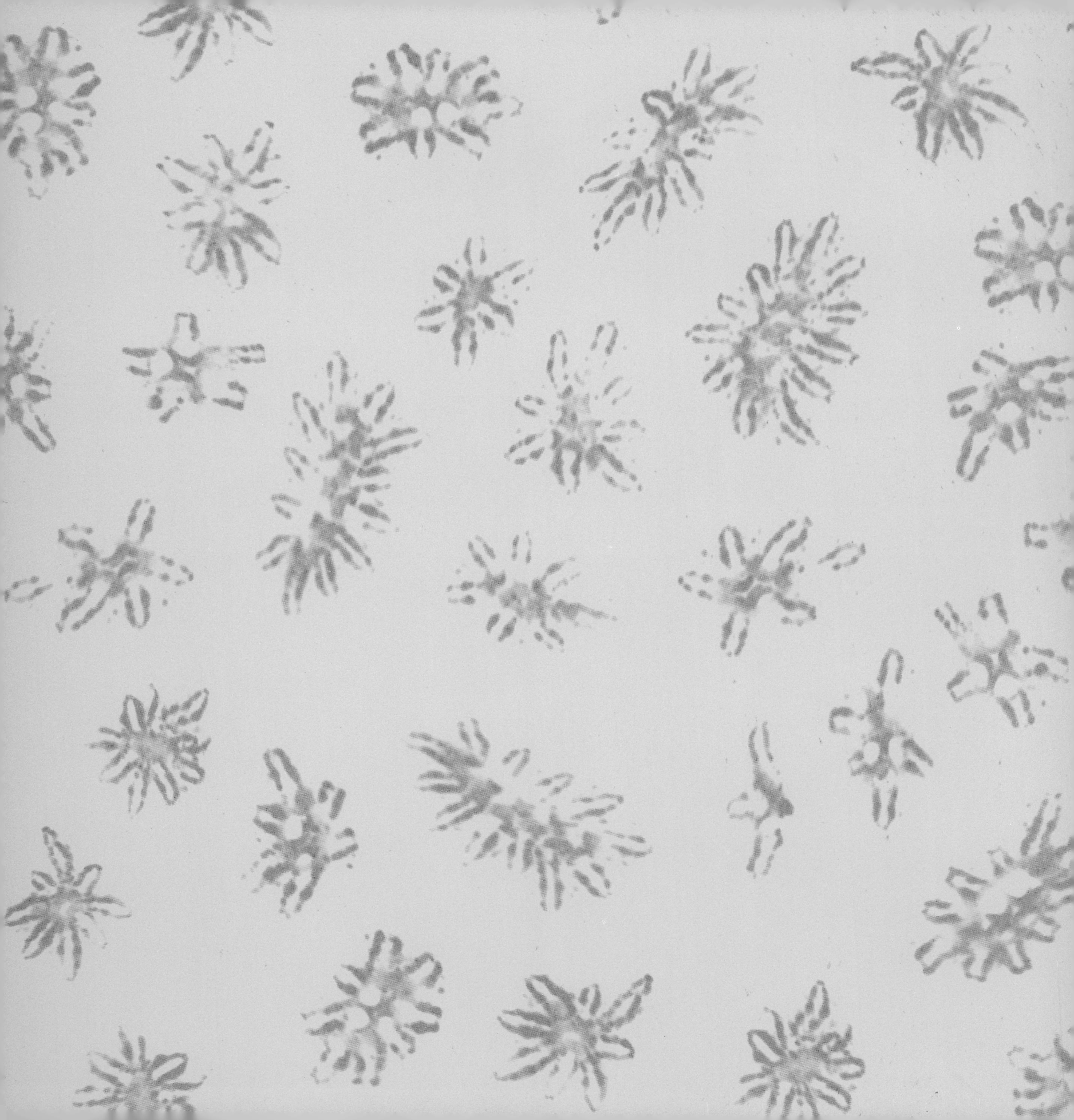

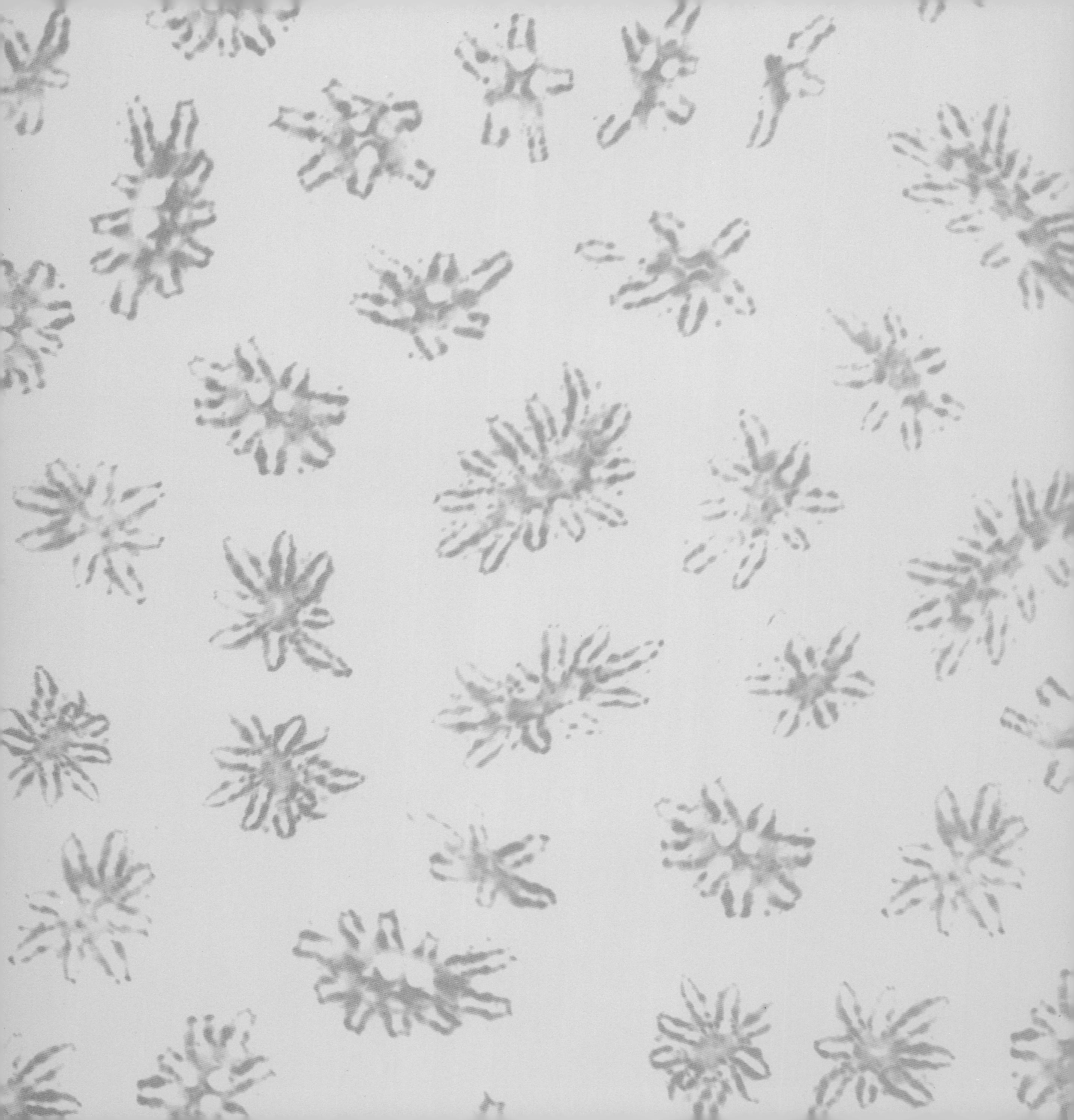